MW01503890

STUDY GUIDE FOR

GENERAL CHEMISTRY: PRINCIPLES AND STRUCTURE
Second Edition

James E. Brady
St. John's University,
New York

SOLUTIONS TO SELECTED NUMERICAL PROBLEMS FROM

GENERAL CHEMISTRY: PRINCIPLES AND STRUCTURE
Second Edition

Theodore W. Sottery
University of Maine
Portland-Gorham

JOHN WILEY & SONS, INC.
New York · Chichester · Brisbane · Toronto

ISBN 0 471 03498 3
Printed in the United States of America

10 9 8 7 6 5 4 3

PREFACE

In this second edition I have retained the same overall format of the Study Guide because so many students have found it useful. Since students generally study a chapter one section at a time, the Study Guide is divided into sections that exactly parallel those in the textbook. Each begins with a statement of objectives that prepares the student for what is to be presented and outlines what he or she is to accomplish. Also included in each section is a brief review. Here the key concepts that are developed in the section are summarized to bring them into focus. In some instances, especially in the critical early chapters, additional explanations of major topics are presented. In this edition I have added additional worked-out examples where they may be of use to complement those that appear in the textbook.

Most sections contain a brief self-test intended to allow students to test their mastery of the subject matter before moving on. Many of these self-tests have been expanded. They generally provide questions of graded difficulty to permit students to progress from simple problems in the direction of more complex ones. None of the problems are very difficult, however, since there is an ample number of difficult problems (those marked with an asterisk) available in the text itself.

Each section concludes with a list of new terms that have appeared in the corresponding section of the text. Students are urged to learn the meanings of these new terms before moving on.

New to the second edition of the Study Guide is the inclusion of detailed solutions to a selected sampling of typical numerical problems that appear in the end-of-chapter problem exercises of the textbook. These should help students to study problem solving in a more independent fashion by providing a path over typical stumbling blocks.

James E. Brady

CONTENTS

BEFORE YOU BEGIN...

Before you begin your general chemistry course, read the next several pages. They're designed to tell you how to use this study guide and to give you a few tips on improving your study habits.

How to Use the Study Guide

This book has been written to parallel the topics covered in your text, General Chemistry: Principles and Structure. For each section in the text you'll find a corresponding section in the study guide. In the study guide the sections are divided into Objectives, Review, Self-Test and New Terms. Before you read a section in the text, read the Objectives in the study guide. This will give you a feeling for what to keep an eye on as you read the text. It should help you understand what you must pay attention to.

After you've read a section, return to the study guide and read the Review. This will point out specific ideas that you should be sure you have learned. Sometimes you will be referred back to the text to review specific items there. Sometimes there will be additional explanations of difficult or important concepts, and in some instances there will be additional worked-out sample problems. Work with the review and the text together to be sure you have mastered the material before going on.

In most sections you will also find a short Self-Test to enable you to test your knowledge and problem-solving ability. The answers to all of the self-test questions are located at the end of the chapters in the study guide. Try to answer the questions without having to look at the answers. A space is left after each question so that you can write in your answers and then check them all after you've finished.

An important aspect of learning chemistry is becoming familiar with the language. There are many cases where lack of understanding can be traced to a lack of familiarity with some of the terms used in a discussion or a problem. A great deal of effort was made in your textbook to avoid using a term without first adequately defining it. Once it has been defined, however, it's used with the assumption that you've learned its meaning. It's important, therefore, to learn new terms as they appear, and for that reason, most of them are set in boldface type in the text. At the end of each section of the study guide there is a list of New Terms. Look them over and check to be sure you know their meaning before proceeding on to the next section. You might find it worthwhile to write out their meanings in your chemistry notebook. This will help you review them later when you prepare for quizzes or examinations.

Study Habits

You say you want to get an A in chemistry? That's not as impossible as you may have been led to believe, but it is going to take some work. The key is efficient study, so your precious study time isn't wasted. Efficient study requires a regular routine, not hard study one night and nothing the next. At first it is difficult to train yourself, but after a short time your study routine will indeed become a study habit and your chances of success in chemistry, or any other subject, will be greatly improved.

To help you get more out of class, try to devote a few minutes the evening before to reading, in the text, the topics that you will cover the next day. Read the material quickly just to get a feel for what the topics are about. Don't worry if you don't understand everything; the idea at this stage is to be aware of what your teacher will be talking about.

Your lecture instructor and your textbook serve to complement one another; they provide you with two views of the same subject. Try to attend lecture regularly and take notes during class. These should include not only those things your teacher writes on the blackboard, but also the important points he or she makes verbally. If you pay attention carefully to what your teacher is saying in class, your notes will probably be somewhat sketchy. They should, however, give an indication of the major ideas. After class, when you have a few minutes, look over your notes and try to fill in the bare spots while the lecture is still fresh in your mind. This will save you much time later when you finally get around to studying your notes in detail.

In the evening (or whatever part of the day you close yourself off from the rest of the world to really study intensely) review your class notes once again. Use the text and study guide as directed above and really try to learn the material presented to you that day. If you have prepared before class and briefly reviewed the notes afterward, you'll be surprised at how quickly and how well your concentrated study time will progress. You may even find yourself enjoying chemistry!

As you study, continue to fill in the bare spots in your class notes. Write out the definitions of new terms in your notebook. In this way, when it comes time for an exam you should be able to review for it simply from your notes.

At this point you're probably thinking that there isn't enough time to do all the things described above. Actually, the preparation before class and brief review of the notes shortly after class take very little time and will probably save more time than they consume.

Well, you're on your way to an A. There are a few other things that can help you get there. If you possibly can, spend about 30 minutes to an hour at the end of a week to review the week's work. Psychologists have found that a few brief exposures to a subject are more effective at fixing them in the mind than a "cram" session before an exam. The brief time spent at the end of a week can save you hours just before an exam (efficiency!). Try it (you'll like it); it works.

There are some people (you may be one of them) who still have difficulty with chemistry even though they do follow good study habits. Often this is because of weaknesses in their earlier education. If, after following intensive study, you are still fuzzy about something, speak to your teacher about it. Try to clear up these problems before they get worse. Sometimes, by having study sessions with fellow classmates you can help each other over stumbling blocks.

Problem Solving

A stumbling block for many chemistry students is numerical problems. Both the textbook and this study guide have worked-out examples in which the solutions to problems are given in rather great detail. Your instructor will also be showing you how to solve problems. But this is not enough! Learning to solve chemistry problems is like learning to play a musical instrument or drive a car. You only learn by doing. Even if you "understand" how a problem is worked out, you still have to try others yourself to see if you really understand the material sufficiently to solve them. Keep working out problems until you can do them; then you can stop. All the problems in the study guide have answers given. The even-numbered numerical problems in the text also have answers given in Appendix E. Work on these so that you can see whether you are getting them correct. Refer also to the worked-out solutions that appear at the back of the study guide. These show how various problems in the text are solved.

One of the goals of both the text and this study guide is to teach you how to solve numerical problems. Perhaps this one aspect of chemistry, more than all others, makes you fearful about your fate in the course - you're afraid of the "math". Actually, though, there is very little mathematics involved in solving most of the chemistry problems you will meet. Most of the difficulty comes from trying to interpret a question so that you know what kind of problem you're supposed to solve. In this section we'll go over some basic approaches to solving problems. If you have difficulty with a problem later on, review the ideas presented here and try to apply them. You'll find that they're useful not only in chemistry, but in other areas as well, including problems you encounter in day-to-day living.

"Word problems" always seem to present students with the most difficulty. "What am I supposed to find?" "Where do I begin?" These are the kinds of questions you've probably asked yourself when faced with a word problem. Many people have found the following to be the most effective way of approaching problems of this type.

Step 1. First, preview the problem to get an overview of the question - the "big picture". At this point, don't get bogged down by details. Don't worry about numbers or specific formulas that may be encountered in the question. Read the entire question without trying to analyze it in detail. Remember, at this point you're only interested in getting a view of the whole problem.

Step 2. After you've looked over the entire problem, the next step is to identify what it is you are asked to compute. Look for key phrases such as "Find...." or "How many...." or "What is...." These allow you to know where you're headed in the solution. You might also try to make an educated guess at the magnitude of the answer, although this isn't essential at this point.

Step 3. Now that you know where you're headed, look over the information provided in the question. Don't worry about the numbers yet; simply examine the nature of the preliminary data. Sometimes it's helpful to extract the data from a word problem and tabulate it so that it isn't cluttered with words.

Step 4. Consider next the kinds of calculations that you must perform on the data. Don't worry about the numbers yet. Simply analyze how you must combine the data in the problems to get the answer you want. Be sure you have everything you need. If you use the factor-label method described in Appendix C of the text, you should be able to write simple equality statements such as:

$$1 \text{ ft} = 12 \text{ in.}$$
$$1 \text{ yd} = 3 \text{ ft}$$

Notice that these two statements have the units "ft" in common and provide sufficient information to convert yards to inches, or vice versa. Be sure your equality statements connect all the units so that you have a path from the starting data to the final answer.

In this step you also must be sure you have any necessary chemical equations or mathematical formulas. Be sure to write them down on paper - don't try to work with them in your head.

Step 5. Well, now you can finally worry about the numbers! At this point all of the necessary information has been compiled and you've decided how you must solve the problem. Now you should go about inserting numbers into formulas or constructing and applying the conversion factors as described in Appendix C of the text. If you've done your preparation in Steps 1 to 4, obtaining an answer in this step should not be difficult.

Step 6. Take a deep breath, you've done it! The problem is solved. As a final point, look at the answer you obtained. Does it seem reasonable? Are the units correct? If so, you're finished.

Time to Begin

Well, it's hoped that the few suggestions presented in this introduction will help you over the hurdles in chemistry. Move on to the course now, and good luck on getting that A!

1 INTRODUCTION

As its name implies, this chapter is meant to introduce you to the study of chemistry. It begins to lay the foundation for the remainder of the course. If you have had chemistry in high school, perhaps much of the material covered in this chapter will be familiar to you. You can test your knowledge by reviewing the list of new terms at the end of each section in Chapter 1 of the Study Guide and by taking the self-tests below. If you've never taken chemistry before, you should be sure to begin the course properly by gaining a thorough understanding of the topics treated here.

1.1 THE SCIENTIFIC METHOD

Objectives

To undertand how science develops through the process of observation, formation of theories, and the design of new experiments that test these theories. You should know the distinction between a law and a theory; between qualitative and quantitative observations.

Review

The scientific method is the procedure that scientists use, either consciously or unconsciously, in their investigation of nature. Data are collected and condensed into laws. Theories are invented in an attempt to explain the laws. The theories suggest new experiments that produce new data, new laws and ultimately new theories. This cycle repeats itself over and over as our understanding of nature grows.

Self-Test (True or False)

1. A law is based on repeated observation. ______

2. A law is an explanation of a theory. ______

3. Theories can always be proven to be correct. ______

4. Laws are often expressed in the form of a mathematical equation. ______

5. A hypothesis is a tentative law. ______

6. Numbers are usually associated with qualitative observations. ______

New Terms

scientific method	law
qualitative observation	hypothesis
quantitative observation	theory
data	

1.2 MEASUREMENT

Objectives

To understand that the extent of our knowledge of the world about us is limited by the precision of the measurements that we make. You should be able to recognize the number of significant digits in a number and be able to express the result of a computation to the proper number of significant figures.

Review

Remember that in counting up significant figures in a number, only zeros that are not required for the sole purpose of locating the decimal point should be included. Some examples are:

number	number of significant figures
302	3
0.012	2
2.012	4
0.0120	3

In performing computations with numbers that come from measurements, remember these rules:

1. Multiplication or division. The answer has the same number of significant figures as the least precise factor in the calculation; for example,

3.05 x	1.3 =	3.965 = 4.0
(3 sig. figures)	(2 sig. figures)	(answer rounded to 2 sig. figures)

2. Addition or subtraction. The number of significant figures in the answer is controlled by the quantity having the largest uncertainty; for example,

	214.3	(implies uncertainty of $\pm$ 0.1)
this quantity has largest uncertainty →	+ 21	(implies uncertainty of $\pm$ 1)
	235	(answer has implied uncertainty of $\pm$ 1)

Exact numbers come from definitions. For instance, 1 mile is exactly 5280 feet, with no uncertainty. Similarly, 1 foot is exactly equal to 12 inches, no more or less. In calculations these numbers may be assumed to possess any desired number of significant figures.

Example 1.1

A desk was measured to be 34.3 in. along its smallest length. Will it fit through a door that is known to be 2.75 feet wide?

Solution

Let's convert 34.4 in. to feet. We can use the relationship between feet and inches to construct a conversion factor (see Appendix C in the text) that enables us to change the units inches into the units feet.

$$34.3\ \cancel{\text{in.}}\left(\frac{1\ \text{ft}}{12\ \cancel{\text{in.}}}\right) = 2.86\ \text{ft}$$

Notice that we may assume that both the 1 and the 12 have as many significant figures as we wish. Since there are three significant figures in 34.3, the answer can be expressed to three significant figures. Also note that the desk won't fit through the door!

In Example 1.1 we have cancelled units just as on Page 5 of the text. You should spend some time now to study and review the factor-label method. It is described in detail in Appendix C (p 759) of the text. Although this method may seem foreign to you now, if you learn to apply it, you will find that setting up the arithmetic of chemistry problems is really not very difficult at all.

Self-Test

7. Give the number of significant figures in each of the following.

(a) 205.3 ______

(b) 113 ______

(c) 200.0 ______

(d) 0.005 ______

(e) 0.0000700 ______

8. Evaluate the following expressions to the proper number of significant figures (assume all numbers represent measured quantities).

(a) $2.43 \times 1.875 =$ ______

(b) $0.017 \times 5.968 =$ ______

(c) $1.43 \times 2.584 \times 0.008 =$ ______

(d) $12.5 \div 2.8 =$ ______

(e) $14.34 \div 4.780 =$ ______

(f) $5.146 + 0.002 =$ ______

(g) $5.146 + 0.02 =$ ______

(h) $8.08 + 80.8 =$ ______

(i) $14.45 + 7.521 + 100.3 =$ ______

(j) $2.92 - 8.4 =$ ______

New Terms

significant figures
precision
accuracy

1.3 UNITS OF MEASUREMENT

Objectives

To learn the basic SI and metric systems of units and to become familiar with conversions from one unit to another within the metric system. You should also learn to express numbers in scientific notation.

Review

You should familiarize yourself with the basic SI units. The units that you should be sure to learn for this course are:

measurement		unit
length	-	meter
mass	-	gram
volume	-	liter

Larger or smaller units are derived from these using an appropriate prefix:

prefix	multiply by	example
kilo	1000	kilogram (kg) = 1000 grams
centi	1/100	centimeter (cm) = 1/100 meter
milli	1/1,000	millimeter (mm) = 1/1000 meter
micro	1/1,000,000	microgram (μg) = 1/1000000 gram
nano	1/1,000,000,000	nanometer (nm) = 1/1000000000 meter

Learn to convert from one unit to another (e.g., liters to milliliters, centimeters to millimeters, etc.). Remember:

1 cm = 10 mm
1 liter = 1000 ml = 1000 cm^3 (or cc)

Most conversions that you will encounter between the English system and the metric system can be handled by remembering the following:

length:	1.00 inch = 2.54 cm
weight:	2.20 lb = 1.00 kg
volume:	1.00 ounce = 29.6 ml

These can provide the cross-over between the two systems of units. For example, to convert 40 ft into meters, first change ft to in., then in. to cm, and finally cm to meters.

$$4.0\,\cancel{\text{ft}}\left(\frac{12\,\cancel{\text{in.}}}{1\,\cancel{\text{ft}}}\right)\left(\frac{2.54\,\cancel{\text{cm}}}{1\,\cancel{\text{in.}}}\right)\left(\frac{1\text{ m}}{100\,\cancel{\text{cm}}}\right) = 1.2\text{ m}$$

Note the cancellation of units. In the Self-Test, practice using units to guide the arithmetic. Remember, if the units don't cancel properly you will obtain the wrong answer, no matter how much you paid for your calculator!

Review, in Appendix C (Page 760), the writing of numbers in scientific notation. Even though your calculator may handle scientific notation, you should understand the principles of arithmetic operations on numbers expressed in this form. See the following Self-Test for practice.

Self-Test (fill in the blanks)

9. Make the following conversions:

(a) 150 m = __________ cm

(b) 27 cm = __________ mm

(c) 1.50 liter = __________ ml

(d) 0.002 g = __________ μg

(e) 100 cm^2 = __________ mm^2

(f) 253 ml = __________ liter

(g) 0.143 g = __________ mg

(h) 1 m^3 = __________ cm^3

(i) 1 km = __________ cm

(j) 84 ml = __________ liter

10. Make the following conversions:

(a) 12.4 in. = __________ cm

(b) 18.3 cm = __________ in.

(c) 1.2 mm = __________ in.

(d) 18.0 m = __________ yd

(e) 1.3 ft^2 = __________ m^2

(f) 1400 cm^3 = __________ qt

(g) 185 km = __________ miles

(h) 1.00 lb = __________ g

(i) 84.0 g = __________ oz

(j) 1.00 ton = __________ kg

11. Without using a calculator, express the following in scientific notation:

(a) 1,400 = __________

(b) 275.3 = __________

(c) 0.00307 = __________

(d) 0.00002 = __________

12. Without using a calculator, write the following in ordinary decimal notation:

(a) 3.0×10^3 = 3000

(b) 2.0×10^6 = ________

(c) 1.5×10^{-5} = ________

(d) 34×10^{-7} = ________

(e) 0.025×10^3 = ________

New Terms

metric system	liter
gram	exponential notation
meter	scientific notation

1.4 MATTER

Objectives

To learn the distinction between weight and mass. You should learn the definition of the term, matter.

Review

Matter has mass and occupies space. The quantity of matter in an object is specified by giving its mass. Mass is the object's resistance to a change in velocity; weight is the force with which an object is attracted to the earth. Mass is measured by comparing the weights of objects on a balance.

New Terms

matter	weight
mass	balance

1.5 PROPERTIES OF MATTER

Objectives

To distinguish between intensive and extensive properties; you should learn the meaning of density. You should understand the difference between physical properties and chemical properties of matter.

Review

Remember, an intensive property is one that is independent of the size of the sample under consideration. For instance, all samples of pure water, regardless of size, freeze at $32^{o}F$; freezing point is an example of an intensive property. The volumes occupied by different samples of water are different, depending on the amount of water in each sample.

Melting point and volume are also physical properties; that is, they may be specified without referring to another substance. Chemical properties are always described by relating one substance to another.

Density is a very useful intensive property. It relates mass to volume. It can be used to calculate the volume of a given mass of substance, and vice versa. The three quantities, density (d), mass (m), and volume (V) are related by the equation,

$$d = \frac{m}{V} \tag{1.1}$$

If you know any two quantities in this equation, you can calculate the third.

Example 1.2

What volume would be occupied by 8.53 g of a substance whose density is 2.54 g/ml?

Solution

You can solve this problem either by Equation 1.1 above or by the cancellation of units. Solving Equation 1.1 for the volume,

$$V = \frac{m}{d}$$

and substituting

$$V = \frac{8.53 \text{ g}}{2.54 \text{ g/ml}}$$

gives

$$V = 3.36 \text{ ml}$$

To solve the problem by unit cancellation you must realize that the density is a conversion factor relating mass and volume.

$$d = \frac{2.54\ \text{g}}{1\ \text{ml}} \qquad \text{or} \qquad \frac{1}{d} = \frac{1\ \text{ml}}{2.54\ \text{g}}$$

Therefore,

$$8.53\ \cancel{\text{g}}\left(\frac{1\ \text{ml}}{2.54\ \cancel{\text{g}}}\right) = 3.36\ \text{ml}$$

Self-Test

13. Indicate whether each of the following are intensive (I) or extensive (E) properties:

(a) mass ______ (d) boiling point ______

(b) color ______ (e) density ______

(c) volume ______

14. Indicate whether each of the following are physical (P) or chemical (C) properties:

(a) nickel chloride is green. ______

(b) carbon monoxide combines with oxygen to produce carbon dioxide. ______

(c) grain alcohol boils at 78.5°C. ______

(d) ozone (produced in smog) is very reactive toward gasoline vapors. ______

(e) carbohydrates are metabolized in the body to produce carbon dioxide and water. ______

15. An object with a mass of 14.3 g displaces 5.22 ml of water when placed in a graduated cylinder. Calculate the density of the object.

______

16. What is the mass of 25.0 ml of an oil if its density is 0.843 g/ml?

______

17. What volume of alcohol (density = 0.789 g/ml) has a mass of 18.0 g?

______

New Terms

extensive properties
intensive properties
density
physical properties
chemical properties

1.6 ELEMENTS, COMPOUNDS, AND MIXTURES

Objectives

To understand the distinction between these three classes of substances.

Review

Remember, mixtures may be of variable composition, such as solutions of salt in water. Compounds and elements are always of fixed (constant) composition. Mixtures may be separated by physical means into their component compounds. Compounds can only be separated into elements by chemical reaction. In order of decreasing complexity we have: mixtures, compounds, elements. Elements are the simplest substances that are encountered in the chemistry laboratory. Study Figure 1.8 on Page 14 of the text.

New Terms

element
compound
mixture
distillation
chromatography
thin-layer chromatography
phase
homogeneous
heterogeneous

1.7 THE LAWS OF CONSERVATION OF MASS AND DEFINITE PROPORTIONS

Objectives

To appreciate the historical significance of these two important chemical laws.

Review

These two basic laws of chemistry govern much of our quantitative thinking about chemical reactions.

New Terms

law of conservation of mass
law of definite proportions
law of definite composition

1.8 THE ATOMIC THEORY OF DALTON

Objectives

To appreciate the historical significance of the development of the atomic theory. You should understand how Dalton's theory explains the chemical laws in Section 1.7 and how it predicts the law of multiple proportions.

Review

Review the definition of a molecule. Below is a sample problem dealing with the law of multiple proportions.

Example 1.3

Two compounds are formed between copper and oxygen. In one there is 0.290 g of oxygen combined with 2.30 g of copper; in the other there is 0.466 g of oxygen combined with 1.85 g of copper. Show that these data demonstrate the law of multiple proportions.

Solution

You must calculate the weight of one element (let's say oxygen) combined with the same weight of the other element (copper) in the two compounds. The data supplied gives you a chemical equivalence between copper and oxygen in the two compounds.

Compound I	2.30 g copper ∼ 0.290 g oxygen
Compound II	1.85 g copper ∼ 0.466 g oxygen

The weight of oxygen that would be combined with 1.00 g of copper in each of these compounds is:

Compound I

$$1.00\ \cancel{\text{g copper}}\left(\frac{0.290\text{ g oxygen}}{2.30\ \cancel{\text{g copper}}}\right) = 0.126\text{ g oxygen}$$

Compound II

$$1.00\ \cancel{\text{g copper}}\left(\frac{0.466\text{ g oxygen}}{1.85\ \cancel{\text{g copper}}}\right) = 0.252\text{ g oxygen}$$

The law of multiple proportions holds if the ratio of these weights of oxygen is a ratio of small whole numbers. Therefore, you set up the ratio,

$$\frac{0.126\text{ g oxygen}}{0.252\text{ g oxygen}} = \frac{1}{2}$$

Self-Test

18. In two compounds, each containing 1.00 g of carbon, there are 0.333 g and 0.167 g of hydrogen, respectively. What is the ratio of weights of hydrogen in the two compounds?

____

19. Phosphorus and oxygen form two compounds. In compound I there are 3.47 g of oxygen combined with 2.68 g of phosphorus; in the other there are 2.82 g of oxygen combined with 3.64 g of phosphorus. What is the ratio of the weights of oxygen that combine with 1.00 g of phosphorus in the two compounds?

____

20. A 4.00-g sample of cupric bromide was heated, driving off some of the bromine and leaving 2.57 g of cuprous bromide. This cuprous bromide was then decomposed to give bromine and pure copper (1.14 g). What is the ratio of the weights of bromine in the two copper compounds?

____

New Terms

molecule
law of multiple proportions

1.9 ATOMIC WEIGHTS

Objectives

To understand how a table of atomic weights can be established by comparing the relative weights of the elements that combine to form compounds of known composition.

Review

The atomic weights that we use are not the weights of individual atoms, but rather the relative weights of atoms compared to one particular isotope of carbon as a standard. The atomic mass of carbon-12 is exactly 12 amu.

New Terms

isotopes

1.10 SYMBOLS, FORMULAS, AND EQUATIONS

Objectives

To begin to become familiar with chemical symbols, formulas, and chemical equations.

Review

Note that in writing the symbol for an element the first letter is capitalized while the second is not. The subscripts in a formula denote the relative numbers of each kind of element contained in that compound. For example, $Na_2S_4O_7$ contains two atoms of sodium (Na), four atoms of sulfur (S) and seven atoms of oxygen (O). The formula $Ca(NO_3)_2$ shows one calcium atom, two nitrogen atoms and six oxygen atoms. The formulas for hydrates show the numbers of water molecules trapped in crystals. For example, $CrCl_3 \cdot 6H_2O$ contains six water molecules for each $CrCl_3$.

Chemical equations are used to indicate what occurs during a chemical reaction. An equation is balanced if there are the same number of atoms of each element on the reactant side (left side) of the arrow as there are on the product side (right side). The symbols s = solid, l = liquid, g = gas, and aq = aqueous solution are used sometimes to indicate the phases of reactants and products.

Self-Test

Refer to the alphabetical list of elements on the inside front cover of the textbook to check your answers to Questions 21 and 22.

21. Write the symbol for each of the following elements.

(a) sodium	________	(f) copper	________
(b) sulfur	________	(g) chlorine	________
(c) oxygen	________	(h) potassium	________
(d) hydrogen	________	(i) magnesium	________
(e) iron	________	(j) carbon	________

22. What are the names of the following elements?

(a) Sn ________________	(d) O ________________
(b) Br ________________	(e) P ________________
(c) Al ________________	(f) Cr ________________

(g) Ag ________________ (i) I ________________

(h) As ________________ (j) He ________________

23. Indicate the number of atoms of each type in the formulas below.

(a) KCl ________________

(b) NO_2 ________________

(c) N_2O_4 ________________

(d) $(NH_4)_3PO_4$ ________________

(e) $Al_2(SO_4)_3$ ________________

(f) $KAl(SO_4)_2 \cdot 12H_2O$ ________________

24. Which of the equations below are not balanced?

(a) $CaO + H_2O \longrightarrow Ca(OH)_2$ ________

(b) $CaCl_2 + H_2SO_4 \longrightarrow HCl + CaSO_4$ ________

(c) $Br_2 + 2NaOH \longrightarrow NaOBr + NaBr + H_2O$ ________

(d) $SO_2 + O_2 \longrightarrow SO_3$ ________

(e) $Al_2(SO_4)_3 + 2CaCl_2 \longrightarrow 2BaSO_4 + AlCl_3$ ________

New Terms

chemical symbol
chemical formula
chemical equation
hydrate
reactants
products
coefficients
balanced equation

1.11 ENERGY

Objectives

To learn the difference between kinetic energy and potential energy. You should understand the difference between heat and temperature and the units used to express them. You should also learn the concept of specific heat.

Review

Kinetic energy is associated with motion ($KE = 1/2\ mv^2$); potential energy is associated with the distance of separation between particles that either attract or repel one another. Remember that if there is neither an attraction nor repulsion between two objects, there are no potential energy changes when the objects move toward or away from each other.

Temperature is an intensive quantity that determines the direction of heat flow (hot → cold). The Celsius scale (or centigrade scale) defines 0^oC as the melting point of ice (which is the same as the freezing point of water). The boiling point of water is defined as 100^oC. The kelvin scale has -273^oC as its zero point. Remember that the size of the degree unit is the same on the Celsius and kelvin scales. A temperature change of one Celsius degree is the same as a change of one kelvin degree. You should be able to convert from one of these temperature scales to another.

Various units used to express energy are the erg, joule (J), kilojoule (kJ), calorie (cal), and kilocalorie (kcal). Remember these conversions:

$$1 \text{ cal} = 4.1840 \text{ J}$$
$$1 \text{ kcal} = 4.1840 \text{ kJ}$$

Review the definition of specific heat and Example 1.5 on Page 24 of the text.

Self-Test

25. Perform the following conversions:

(a) 25^oC = ______________ K

(b) -30^oC = ______________ K

(c) 350 K = ______________ ^oC

(d) 77 K = ______________ ^oC

(e) 4.0^oC = ______________ ^oF

(f) 50^oF = ______________ ^oC

(g) 254 kJ = ______________ kcal

(h) 32.5 kcal = ______________ kJ

26. A 1.50-g piece of gold absorbed 0.162 cal when its temperature was raised by 3.50^oC. Calculate the specific heat of gold in the units.

(a) cal/g^oC ______________

(b) J/g^oC ______________

<u>New Terms</u>

kinetic energy
potential energy
exothermic
endothermic
law of conservation of energy
erg
joule
temperature
Fahrenheit scale
Celsius scale
kelvin scale
calorie
kilocalorie
specific heat

ANSWERS TO SELF-TEST QUESTIONS

1. true 2. false 3. false 4. true 5. false 6. false 7. (a) 4 (b) 3 (c) 4 (d) 1 (e) 3 8. (a) 4.56 (b) 0.10 (c) 0.03 (d) 4.5 (e) 3.000 (f) 5.148 (g) 5.17 (h) 88.9 (i) 122.3 (j) -5.5 9. (a) 15000 cm (b) 270 mm (c) 1500 ml (d) 2000 μg (e) 10,000 mm^2 (f) 0.250 liter (g) 143 mg (h) 1,000,000 cm^3 (i) 100,000 cm 10. (a) 31.5 cm (b) 7.20 in. (c) 0.047 in. (d) 19.7 yd (e) 0.12 m^2 (f) 1.5 qt (to 2 sig. fig.) (g) 115 mi (h) 455 g (i) 2.96 oz (j) 909 kg 11. (a) 1.4×10^3 (b) 2.753×10^2 (c) 3.07×10^{-3} (d) 2×10^{-5} 12. (b) 2,000,000 (c) 0.000015 (d) 0.0000034 (e) 25 13. (a) E (b) I (c) E (d) I (e) I 14. (a) P (b) C (c) P (d) C (e) C 15. 2.74 g/ml 16. 21.1 g 17. 22.8 ml 18. 2 to 1 19. g O in I/g O in II = 1.29/0.775 = 1.66 = 5/3 20. 2.86/1.43 = 2/1 21. (a) Na (b) S (c) O (d) H (e) Fe (f) Cu (g) Cl (h) K (i) Mg (j) C 22. (a) tin (b) bromine (c) aluminum (d) oxygen (e) phosphorus (f) chromium (g) silver (h) arsenic (i) iodine (j) helium 23. (a) K, 1; Cl, 1 (b) N, 1; O, 2 (c) N, 2; O, 4 (d) N, 3; H, 12; P, 1; O, 4 (e) Al, 2; S, 3; O, 12 (f) K, 1; Al, 1; S, 2; O, 20; H, 24 24. (a) balanced (b) unbalanced (c) balanced (d) unbalanced (e) unbalanced 25. (a) 298 K (b) 243 K (c) 77°C (d) -196°C (e) 39.2°F (f) 10°C (g) 60.7 kcal (h) 136 kJ 26. (a) 0.0309 cal/g$^\circ$C (b) 0.129 J/g$^\circ$C

2 STOICHIOMETRY: CHEMICAL ARITHMETIC

This chapter deals with calculations involving quantities of chemical substances, either combined together in a compound or reacting with one another in a chemical reaction. It is very important that you thoroughly understand the concepts developed here because they are necessary in future discussions in other chapters. This is particularly true of the mole concept discussed in Section 2.1. Most students who develop difficulties with chemistry have not really acquired a genuine "feel" for the mole. Therefore, this is really a very important chapter for you to master well. A little extra time spent here may save you a lot of grief later.

2.1 THE MOLE

Objectives

To learn to think of the mole as the "chemist's dozen". You should develop the ability to translate between ratios of numbers of atoms and molecules that combine and ratios of numbers of moles of these substances that combine. You should also be able to convert from moles to grams, and vice versa. You should learn to use Avogadro's number where appropriate.

Review

Like the dozen (12) or the gross (144), the mole represents a fixed number of objects. These can be atoms, molecules, or anything we wish to consider. One mole of an element or compound contains a large enough quantity of atoms or molecules to be worked with in a laboratory. The most important feature of the mole concept, however, is that in a chemical reaction (for instance, the formation of a compound from its elements) the RATIO in which atoms combine is exactly the same as the RATIO in which moles of atoms combine. When $CrCl_3$ is formed, three Cl atoms are required for each Cr atom. In laboratory-sized quantities, three moles of Cl atoms are required for each one mole of Cr atoms.

A concept that many students find difficult to grasp is that we obtain one mole of $CrCl_3$ from three moles of Cl and 1 mole of Cr.

$$1 \text{ mole Cr} + 3 \text{ moles Cl} \longrightarrow 1 \text{ mole } CrCl_3$$

If this troubles you, consider this analogy:

$$1 \text{ doz frames} + 3 \text{ doz wheels} \longrightarrow 1 \text{ doz tricycles}$$

Both equations involve precisely the same kind of reasoning.

You should learn to convert between laboratory units of grams and chemical units of moles. Remember that one mole of any element has a mass in grams numerically equal to the element's atomic weight. For example, the atomic weight of fluorine is 19.0; 1 mole of F = 19.0 g.

Avogadro's number (6.023×10^{23} things = 1 mole things) is used only when you must translate sizes (mass, volume, length, etc.) between the large world that we work in (with balances, graduated cylinders, and rulers) to the submicroscopic world of individual atoms and molecules.

Self-Test

1. How many moles of F must react with one mole of S to form:

   (a) SF_2 __________ (c) SF_6 __________

   (b) SF_4 __________

2. What mole ratio of carbon to hydrogen is found in propane, C_3H_8?

   ____

3. How many moles of Cl must react with 0.50 mole of P to form PCl_5?

   ____

4. How many moles of oxygen atoms are there in 1.20 moles of Fe_3O_4?

   ____

5. How many moles of iron (Fe) atoms are there in 1.20 moles of Fe_3O_4?

   ____

6. Oxygen occurs as molecules of O_2. How many moles of O atoms are there in 1.40 moles of O_2?

____

7. How many moles of O_2 would be needed to prepare 5.0 moles of N_2O_4?

____

8. How many moles of Cl_2 would be needed to prepare 3.0 moles of PCl_5?

____

9. What is the weight of one mole of:

(a) carbon atoms _______________

(b) potassium atoms _______________

(c) calcium atoms _______________

(d) nickel atoms _______________

(e) bromine atoms _______________

10. How many moles of atoms are there in:

(a) 32.1 g S _______________

(b) 46.0 g Na _______________

(c) 12.5 g Ag _______________

(d) 3.50 g N _______________

11. How many grams does each of the following weigh?

(a) 1 mole Mn _______________

(b) 0.455 mole Al _______________

(c) 1.34 mole Ba _______________

(d) 2.14 mole Zn _______________

12. How many grams of each element are present in 0.250 mole Al_2O_3?

____

13. How many grams of O are needed to react with 0.300 mole S to form SO_3?

____

14. How many grams of Cl must react with 10.0 g C to form C_2Cl_4?

____

15. How many atoms are there in:

(a) 1.00 mole S ______________

(b) 1.00 mole O_2 ______________

(c) 0.341 mole P ______________

(d) 1.85 mole Cl_2 ______________

16. What is the weight in grams of 1 atom of Si? ______________

17. What is the weight in grams of 1 molecule of O_2? ______________

18. How many atoms are there in 5.27 g of Na? ______________

New Terms

mole

2.2 MOLECULAR WEIGHTS AND FORMULA WEIGHTS

Objectives

To learn to calculate molecular weights and formula weights.

Review

The molecular weight, or formula weight, is simply the sum of all of the atomic weights of all of the atoms in the formula of a compound. For example, the molecular weight of glucose, $C_6H_{12}O_6$, is:

carbon	6 x 12.01	=	72.06
hydrogen	12 x 1.008	=	12.10
oxygen	6 x 16.00	=	96.00
formula weight of $C_6H_{12}O_6$		=	180.16

Remember that 1 mole of a compound = 6.02×10^{23} formula units and has a weight in grams numerically equal to the formula weight. Thus, 1 mole $C_6H_{12}O_6$ = 180.16 g.

Self-Test

19. Calculate the formula weights of:

(a) KNO_3 (potassium nitrate) __________

(b) $CO(NH_2)_2$ (urea) __________

(c) $C_9H_8O_4$ (aspirin) ________

(d) CCl_4 (carbon tetrachloride) ________

(e) NaOCl (bleach) ________

20. How many moles are there in each of the following?

(a) 14.3 g of $NaC_{18}H_{35}O_2$ (soap) ________

(b) 142 g of $(CH_3)_2CO$ (acetone - used in nail polish remover) ________

New Terms

molecular weight	electron
formula weight	ion
formula unit	

2.3 PERCENTAGE COMPOSITION

Objectives

To calculate the percentage composition of a compound from its formula. You should be able to calculate the weight of a given element in a compound given its formula.

Review

The weight percent of an element in a compound is calculated by dividing the weight of that element in the compound by the molecular weight and then multiplying by 100. The percent Cl in CCl_4 is:

$$\begin{aligned}\% \text{ Cl in } CCl_4 &= \left(\frac{\text{wt Cl in } CCl_4}{\text{M.W. } CCl_4}\right) 100 \\ &= \left(\frac{(4)(35.5)}{12.0 + (4)(35.5)}\right) 100 \\ &= \left(\frac{142}{154}\right) 100 \\ &= 92.2\%\end{aligned}$$

To calculate the weight of a given element in a sample of a compound you multiply the weight of the sample by the fraction of the compound that is the desired element. For example, to calculate the weight of chlorine in 85.0 g of CCl_4 we multiply the weight of the sample (85.0 g) by the fraction of CCl_4 that is chlorine, $\frac{142 \text{ g Cl}}{154 \text{ g } CCl_4}$. Note that this fraction is the percent divided by 100.

$$\text{wt Cl in sample} = 85.0 \text{ g } \cancel{CCl_4} \left(\frac{142 \text{ g Cl}}{154 \text{ g } \cancel{CCl_4}}\right) = 78.4 \text{ g Cl}$$

Self-Test

21. Calculate the percent compositions of each of the following.

    (a) $NaNO_3$ __________

    (b) $Ca(HCO_3)_2$ __________

    (c) $NiCl_2$ __________

22. Calculate the weight of Ag in 22.0 g of AgCl. __________

23. Calculate the weight of sulfate in 12.3 g of $BaSO_4$. __________

24. In a chemical analysis a 3.14-g sample known to contain $CuSO_4$ and $CuCl_2$ was dissolved in water and treated with $Ba(NO_3)_2$. Solid $BaSO_4$ was formed which was filtered from the solution, dried and weighed. The $BaSO_4$ weighed 2.58 g.

    (a) What weight of SO_4 was in the $BaSO_4$? __________

    (b) What was the weight percent SO_4 in the original sample?

    ____

New Terms

percentage composition

2.4 CHEMICAL FORMULAS

Objectives

To learn what types of information different kinds of chemical formulas provide.

Review

The simplest formula only gives the relative numbers of atoms in the compound. The molecular formula also gives the actual number of each element in a molecule of the compound. The structural formula describes the way in which the atoms in a molecule are linked together:

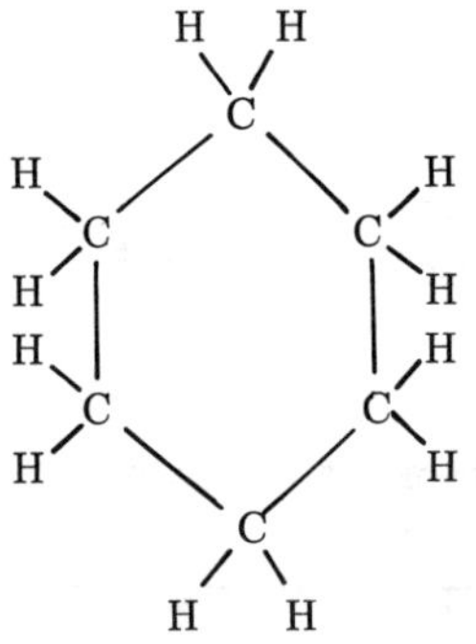

structural formula for cyclohexane

C_6H_{12} — molecular formula for cyclohexane

CH_2 — empirical formula for cyclohexane

Self-Test

12. What is the molecular formula and the empirical formula for the compound whose structural formula is

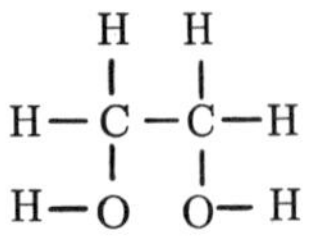

(ethylene glycol - antifreeze)

____

New Terms

simplest formula
empirical formula
molecular formula
structural formula

2.5 EMPIRICAL FORMULAS

Objectives

To calculate empirical formulas from percentage composition or from the weights of elements combined together.

Review

An empirical formula gives the atom ratio of elements in the compound; it also gives the ratio of the number of moles of each element. The atom ratio and mole ratio, of course, have to be identical, based on the definition of the mole. If you are unsure about this, you should review the mole concept.

To determine an empirical formula you must calculate the number of moles of each element combined in a given sample of the compound, as shown in Example 2.10 in the text. The ratio of moles gives the atom ratio. The following is another example:

Example 2.1

A 4.00-g sample of a copper-bromine compound was decomposed, yielding 1.14 g of pure copper. What is the empirical formula of the compound?

Solution

First, we must have the weight of Br combined with the Cu. This can be obtained in this case as the difference between the total weight of compound and the weight of copper in the compound.

$$4.00\text{ g} - 1.14\text{ g} = 2.86\text{ g Br}$$

Thus, there were 2.86 g of Br combined with 1.14 g of Cu in the original sample.

Next, we calculate the number of moles of Cu and Br.

$$1.14\,\cancel{\text{g Cu}}\left(\frac{1\text{ mole Cu}}{63.5\,\cancel{\text{g Cu}}}\right) = 0.0179\text{ mole Cu}$$

$$2.86\,\cancel{\text{g Br}}\left(\frac{1\text{ mole Br}}{79.9\,\cancel{\text{g Br}}}\right) = 0.0358\text{ mole Br}$$

Finally, set up the formula as $Cu_{0.0179}Br_{0.0358}$ and divide through by the smallest subscript to obtain whole numbers.

$$Cu_{\frac{0.0179}{0.0179}}Br_{\frac{0.0358}{0.0179}} = CuBr_2$$

(The data for this question came from Question 20 in Chapter 1 of the Study Guide.)

When an analysis is presented in the form of percent composition by weight, you can write the percentages of each element as weights by assuming 100 g of compound in the sample. Thus 100 g of a compound that is 50% sulfur and 50% oxygen would contain 50 g S and 50 g O.

Self-Test

25. Determine the empirical formulas for the following compounds.

(a) 7.86 g potassium, 7.14 g chlorine ________

(b) 37.9 g Na, 17.1 g P ________

(c) 31.9% K, 29.0% Cl, 39.2% O ________

(d) 69.6% Mn, 30.4% O ________

New Terms

none

2.6 MOLECULAR FORMULAS

Objectives

To determine the molecular formula of a substance, given its empirical formula and its molecular weight.

Review

The molecular formula is always a multiple of the empirical formula. For example, benzene has a molecular formula of C_6H_6; its empirical formula is CH. The empirical formula occurs "six times" in the molecular formula; that is, the subscripts in the empirical formula of benzene must each be multiplied by six to give the molecular formula. The molecular weight of C_6H_6 is 78; the formula weight of CH is 13. The number of times the empirical formula is repeated is obtained by dividing the molecular weight by the empirical formula weight.

$$\frac{78}{13} = 6$$

Self-Test

26. Determine molecular formulas for the following compounds.

empirical formula	molecular weight	molecular formula
CH_2O	180	________
CH_3	30	________
P_2O_3	220	________

HgCl	472.2	________

New Terms

none

2.7 BALANCING CHEMICAL EQUATIONS

Objectives

To write balanced chemical equations.

Review

You should keep in mind the advice that writing a balanced equation is a two-step process. First, write the unbalanced equation with correct formulas for each of the reactants and products. Then, balance the equation. At this point in the course you must balance the equation by inspection; that is, you juggle the coefficients (the numbers preceding the formulas) to make the number of atoms of each kind the same on both sides of the equation. Remember, once you have written correct formulas for the reactants and products you do not change the subscripts in the formulas. Also remember that a properly balanced equation has the smallest whole number set of coefficients. Practice balancing the equations in the Self-Test below.

Self-Test

27. Balance the following equations.

(a) $CuSO_4 + Al \longrightarrow Al_2(SO_4)_3 + Cu$ ________

(b) $KClO_3 \longrightarrow KCl + O_2$ ________

(c) $CO(NH_2)_2 + H_2O \longrightarrow CO_2 + NH_3$ ________

(d) $PCl_5 + H_2O \longrightarrow H_3PO_4 + HCl$ ________

(e) $N_2O_5 + H_2O \longrightarrow HNO_3$ ________

New Terms

none

2.8 CALCULATIONS BASED ON CHEMICAL EQUATIONS

Objectives

To use a balanced chemical equation to perform calculations involving quantities of substances entering into chemical reaction.

Review

A chemical equation such as

$$2H_2 + O_2 \longrightarrow 2H_2O$$

tells us about reacting molecules. It gives information about what happens on a submicroscopic, atomic scale. The mole concept allows us to expand this information up to laboratory-size quantities. Whatever ratio exists between atoms or molecules of reactants and products, the same ratio exists between moles of reactants and products. For example, the equation above tells us that for every two molecules of H_2 that react, one molecule of O_2 will also react. Scaling this to laboratory-sized quantities, we can say that for every two moles of H_2 that react, one mole of O_2 will react.

In dealing with chemical equations, chemists do their thinking in terms of moles. Let's use the equation above as an example. If a chemist knew that he had 0.40 mole of O_2, he would look at the equation and realize immediately that he would require 0.80 mole of H_2. The equation tells him that twice as many moles of H_2 must react as O_2. Similarly, he would also conclude that he would be able to obtain 0.80 mole of H_2O. The equation tells him that however moles of H_2 are consumed, the same number of moles of H_2O will be formed.

The purpose of doing these calculations is to be able to measure proper quantities of reactants (or products) in the laboratory. However, we cannot measure moles directly; instead we can only measure mass (i.e., grams). We therefore must translate back and forth between laboratory units (grams) and chemical units (moles).

There are a number of worked-out examples presented in the text illustrating the types of calculations you might encounter. Let's look at another.

Example 2.2

How many grams of H_2 are needed to react completely with 4.75 g of Fe_2O_3 according to the equation,

$$Fe_2O_3 + 3H_2 \longrightarrow 2Fe + 3H_2O$$

Solution

All reasoning involving H_2 and Fe_2O_3 must take place in chemical units. The solution of this problem can be diagrammed as shown below.

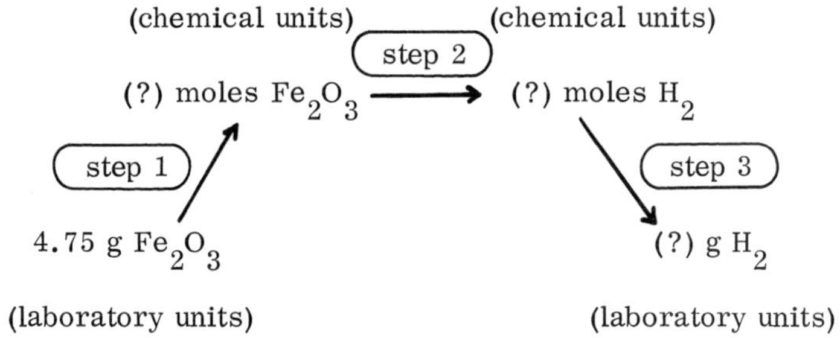

Steps 1 and 3 involve translation between grams and moles (this was covered in Sections 2.1 and 2.2).

Step 2 requires the use of the coefficients in the balanced equation.

Step 1 - Translation

$$4.75 \text{ g } Fe_2O_3 \left(\frac{1 \text{ mole } Fe_2O_3}{159.6 \text{ g } Fe_2O_3}\right) = 0.0298 \text{ mole } Fe_2O_3$$

Step 2 - The coefficients in the equation allow us to establish the chemical equivalency,

$$1 \text{ mole } Fe_2O_3 \sim 3 \text{ moles } H_2$$

This is then used to construct a conversion factor so that we can calculate the number of moles of H_2 required.

$$0.0298 \text{ mole } Fe_2O_3 \left(\frac{3 \text{ moles } H_2}{1 \text{ mole } Fe_2O_3}\right) = 0.0894 \text{ mole } H_2$$

Step 3 - Translation

$$0.0894 \text{ mole } H_2 \left(\frac{2.02 \text{ g } H_2}{1 \text{ mole } H_2}\right) = 0.180 \text{ g } H_2$$

Self-Test

28. The reaction between hydrazine, N_2H_4, and hydrogen peroxide, H_2O_2, has been used to power rockets.

$$N_2H_4 + 2H_2O_2 \longrightarrow N_2 + 4H_2O$$

(a) How many moles of N_2H_4 are required to react with 8.00 moles of H_2O_2? __________

(b) How many moles of N_2 will be formed from 8.00 moles of H_2O_2? __________

(c) How many moles of water will be formed from 8.00 moles of H_2O_2? __________

(d) How many grams of water will be formed when 3.00 moles of N_2H_4 react? __________

(e) How many moles of N_2 will be formed when 500 g of H_2O_2 react? __________

(f) How many grams of H_2O_2 are required to react with 1000 g of N_2H_4? __________

New Terms

none

2.9 LIMITING-REACTANT CALCULATIONS

Objectives

To calculate the amount of products formed when arbitrary amounts of reactants are mixed.

Review

These calculations deal with chemical reactions in which substances are simply mixed together without prior regard for maintaining the proper mole ratios between reactants. In these cases, all reactants usually are not consumed completely; one or more of them remains in excess. The amount of product formed in these situations is controlled by the reactant that is completely used up (the limiting reactant), since once it is gone no more product is able to form. In this type of problem, you first determine the limiting reactant and then base your calculation of the amount of product formed on the amount of the limiting reactant available.

Example 2.3

Zinc and oxygen combine to produce zinc oxide according to the equation,

$$2Zn + O_2 \longrightarrow 2ZnO$$

How much ZnO will be formed if 14.3 g of Zn are mixed with 3.72 g of O_2?

Solution

First, we calculate how many moles of Zn and O_2 are in the mixture.

$$14.3 \text{ g Zn} \left(\frac{1 \text{ mole Zn}}{65.4 \text{ g Zn}}\right) = 0.219 \text{ mole Zn}$$

$$3.72 \text{ g } O_2 \left(\frac{1 \text{ mole } O_2}{32.0 \text{ g } O_2}\right) = 0.116 \text{ mole } O_2$$

Next, we determine the limiting reactant by choosing one reactant and calculating the amount of the other required to give complete reaction. It doesn't matter which we choose; so let's work with the Zn.

$$0.219 \text{ mole Zn} \left(\frac{1 \text{ mole } O_2}{2 \text{ moles Zn}}\right) \sim 0.109 \text{ mole } O_2 \text{ required to react with all the Zn}$$

Notice that we have more O_2 than we need. This means that O_2 will be left over and all of the Zn will react; zinc is the limiting reactant.

Once the limiting reactant is established we use it to calculate the amount of product that will be formed.

$$0.219 \text{ mole Zn} \left(\frac{1 \text{ mole ZnO}}{1 \text{ mole Zn}}\right)\left(\frac{81.4 \text{ g ZnO}}{1 \text{ mole ZnO}}\right) \sim 17.8 \text{ g ZnO}$$

The weight of ZnO formed is 17.8 g.

Self-Test

29. Based on the equation,

$$N_2H_4 + 2H_2O_2 \longrightarrow N_2 + 4H_2O$$

(a) How many moles of N_2 will be formed if 600 g of N_2H_4 are mixed with 1200 g of H_2O_2? ____________

(b) How many grams of H_2O will be produced if 83.5 g of N_2H_4 are mixed with 175 g of H_2O_2? ____________

New Terms

limiting reactant

2.10 THEORETICAL YIELD AND PERCENTAGE YIELD

Objectives

To see that not all reactions produce the theoretical maximum amount of product. You should learn the definitions of theoretical yield and percentage yield.

Review

The theoretical yield is the amount of product that would be produced if the reactants were to combine to the maximum extent possible. We calculate the theoretical yield from the limiting reactant following the procedure in the last section. In simpler cases we calculate it as the maximum amount of product formed from a particular reactant as in Section 2.8. The percentage yield is calculated as shown on Page 44 of the text.

Self-Test

30. Glucose, $C_6H_{12}O_6$, is converted to ethyl alcohol, C_2H_5OH, and CO_2 by fermentation,

$$C_6H_{12}O_6 \longrightarrow 2C_2H_5OH + 2CO_2$$

Starting with 200 g of glucose,

(a) What is the theoretical yield of ethyl alcohol? ____________

(b) If 97.3 g of C_2H_5OH was obtained, what was the percentage yield? ____________

New Terms

theoretical yield
percentage yield

ANSWERS TO SELF-TEST QUESTIONS

1. (a) 2 (b) 4 (c) 6 2. 3/8 3. 2.5 moles Cl 4. 4.80 moles O 5. 3.60 moles Fe 6. 2.80 moles O 7. 10 moles O_2 8. 7.5 moles Cl_2 9. (a) 12.0 g (b) 39.1 g (c) 40.1 g (d) 58.7 g (e) 79.9 g 10. (a) 1.00 mole S (b) 2.00 moles Na (c) 0.116 mole Ag (d) 0.250 mole N 11. (a) 54.9 g Mn (b) 12.3 g Al (c) 184 g Ba (d) 140 g Zn 12. 13.5 g Al, 12.0 g O 13. 14.4 g O 14. 59.2 g Cl 15. (a) 6.02×10^{23} (b) 1.20×10^{24} (c) 2.05×10^{23} (d) 2.23×10^{24} 16. 4.67×10^{-23} g 17. 5.32×10^{-23} g 18. 1.38×10^{23} atoms 19. (a) 101.1 (b) 60.0 (c) 180 (d) 154 (e) 74.5 20. (a) 0.0467 (b) 2.45 21. (a) 27.1% Na, 16.5% N, 56.5% O (b) 24.7% Ca, 1.24% H, 14.8% C, 59.2% O (c) 45.3% Ni, 54.7% Cl 22. 16.6 g Ag 23. 5.06 g SO_4 24. (a) 1.06 g SO_4 (b) (1.06/3.14) x 100 = 33.8% SO_4 25. molecular formula = $C_2H_6O_2$, empirical formula = CH_3O 26. $C_6H_{12}O_6$, C_2H_6, P_4O_6, Hg_2Cl_2 27. (a) $3CuSO_4 + 2Al \longrightarrow Al_2(SO_4)_3 + 3\ Cu$ (b) $2KClO_3 \longrightarrow 2KCl + 3\ O_2$ (c) $CO(NH_2)_2 + H_2O \longrightarrow CO_2 + 2NH_3$ (d) $PCl_5 + 4H_2O \longrightarrow H_3PO_4 + 5HCl$ (e) $N_2O_5 + H_2O \longrightarrow 2HNO_3$ 28. (a) 4.00 moles N_2H_4 (b) 4.00 moles N_2 (c) 16.0 moles H_2O (d) 216 g H_2O (e) 7.35 moles N_2 (f) 2.12×10^3 g H_2O_2 29. (a) 17.6 moles N_2 (H_2O_2 limiting reactant) (b) 185 g H_2O (H_2O_2 limiting reactant) 30. (a) 102 g (b) 95.4%

3 ATOMIC STRUCTURE AND THE PERIODIC TABLE

Atoms are not indivisible particles as Dalton had originally envisioned them. Instead, they are composed of simpler particles, neutrons and positively charged protons in the nucleus of the atom and negatively charged electrons surrounding the nucleus. The first half of this chapter follows the historical developments that have led to the presently accepted theory about atomic structure. The second part of the chapter describes the electronic structure of the atom and how certain observed properties can be related to atomic structure. This is very important because the chemical behavior of atoms is controlled by their electronic structures.

3.1 THE ELECTRICAL NATURE OF MATTER

Objectives

To understand how experimental evidence was accumulated that showed matter to be electrical in nature.

Review

Faraday's experiments on electrolysis showed that chemical reactions could be caused by electricity.

J. J. Thomson's experiments using the cathode ray tube showed electrons to be fundamental particles. He measured the charge-to-mass ratio for the electron.

Self-Test

The first self-test will be found after Section 3.9. Read and study the first nine sections of the chapter, using the Study Guide in the usual way, before attempting to answer the questions in the self-test.

New Terms

electron	cathode rays
gas discharge tube	anode
cathode	coulomb

3.2 THE CHARGE ON THE ELECTRON

Objectives

To understand how the charge on the electron was measured.

Review

R. A. Millikan measured the charge on the electron by measuring the charge on oil drops that had picked up electrons. The charge on the oil drops was always a multiple of -1.60×10^{-19} coulombs, and this value was therefore taken to be equal to the electron's charge. (Your instructor probably doesn't expect you to memorize this number - to be sure, though, you should ask.)

New Terms

none

3.3 POSITIVE PARTICLES, THE MASS SPECTROMETER

Objectives

To understand that atoms must also contain positive particles and that these positive particles are much heavier than the electron. You should learn that charges on particles are expressed in multiples of the charge on the electron.

Review

The mass spectrometer is a device used to measure the charge-to-mass ratio of positive particles (positive ions). These ions always have much smaller e/m ratios than the electron, which tells us that they are much heavier than the electron. The largest e/m ratio is observed for the hydrogen ion which is simply a proton. The proton is a fundamental particle.

The electron is assigned a charge of -1; the proton has a charge of +1. This is because charge is gained or lost by atoms when they gain or lose electrons. A relative charge of -1 really corresponds to an actual charge of -1.60×10^{-19} coulombs. You will almost always deal with relative charges.

New Terms

ion
mass spectrometer
proton

3.4 RADIOACTIVITY

Objectives

To learn the kinds of radioactivity shown by certain kinds of atoms and understand that this phenomenon also shows that there are particles simpler than the atom.

Review

Three basic types of radioactivity are observed:

α-rays composed of He^{2+} ions (α-particles)
β-rays composed of electrons (β-particles)
γ-rays composed of very penetrating radiation similar to X-rays

Since these emissions come spontaneously from atoms of certain substances, these atoms must be constructed of smaller, simpler particles.

New Terms

alpha particle
beta particle
gamma rays
radioactivity

3.5 THE NUCLEAR ATOM

Objectives

To see how experiments led to the idea that the atom has a positive nucleus.

Review

E. Rutherford concluded that the atom must possess a very tiny nucleus containing all of the positive charge in the atom and nearly all its mass. This is the only way he could account for the scattering of α-particles at large angles from thin metal foils.

New Terms

nucleus

3.6 ELECTROMAGNETIC RADIATION

Objectives

To learn about some interrelated properties of light waves.

Review

Remember that light travels at a constant speed (c) through a vacuum, 3.00×10^{10} cm/sec (3.00×10^{8} m/sec). The product of the light's frequency (ν) and wavelength (λ) is equal to c.

$$\lambda \cdot \nu = c$$

The SI unit of frequency is the hertz: 1 Hz = 1 sec^{-1}.

New Terms

electromagnetic radiation	wavelength
frequency	hertz

3.7 X-RAYS AND ATOMIC NUMBER

Objectives

To see that an atom can be identified by its atomic number, which is equal to the number of protons in the atomic nucleus.

Review

The atomic number was originally identified with the frequency of X-rays emitted by an element. Moseley was able to show that the atomic number obtained from experiment was also equal to the number of protons in the nucleus. For example, the atomic number of sodium is found to be eleven; there are eleven protons in the nucleus of a sodium atom.

New Terms

atomic number

3.8 THE NEUTRON

Objectives

To examine the properties of the fundamental particle called the neutron.

Review

Neutrons are particles of zero charge and of mass almost the same as the proton. You should review the properties of the proton, neutron and electron in Table 3.1.

The unit of length, the Ångstrom, is introduced in this section. You should remember that 1 Å $= 10^{-8}$ cm. This is a convenient unit for expressing atomic dimensions.

New Terms

neutron
angstrom

3.9 ISOTOPES

Objectives

To learn the meaning of the term isotope. You should learn how to write the symbol for a given isotope of an element and how to calculate the average atomic mass from the actual isotopic masses and their relative abundances.

Review

Isotopes of the same element have the same atomic number (number of protons) but different numbers of neutrons. Remember that when writing the symbol for an isotope, the atomic number, Z, is a left subscript and the mass number, A (the sum of protons plus neutrons), is a left superscript. The number of neutrons is A - Z. For example,

$$^{70}_{32}Ge \qquad \begin{aligned} Z &= 32 \text{ (32 protons)} \\ A &= 70 \\ A - Z &= 38 \text{ (38 neutrons)} \end{aligned}$$

The relative atomic masses discussed previously are actually average atomic masses. Example 3.2 in the text illustrates how the average atomic mass can be calculated from fractional abundances and accurate relative isotopic masses. This is also shown in the example below.

Example 3.1

Naturally occurring chlorine is composed of a mixture of 75.53% ^{35}Cl, and 24.47% ^{37}Cl. These have isotopic masses of 34.969 and 36.966 amu, respectively. Calculate the average atomic mass of chlorine.

Solution

Multiply the mass of each isotope by its fractional abundance (obtained from percent by dividing by 100). Then add the results to get the average atomic mass.

^{35}Cl	(34.969 amu)(0.7553)	=	26.41
^{37}Cl	(36.966 amu)(0.2447)	=	9.05
	Total	=	35.46

New Terms

mass number

Self-Test on Sections 3.1 to 3.9

Try to answer these questions without referring back to the text or to the review material in this book.

1. (True or False)

(a) Faraday's experiments permitted the determination of the charge-to-mass ratio of the electron. ________

(b) The cathode ray tube used by Thomson is similar to a television picture tube. ________

(c) Cathode rays have different properties for different samples of matter. ________

(d) The charge on the electron was measured by experiments using charged oil droplets. ________

(e) From the data obtained from Thomson's cathode ray tube experiments and Millikan's experiments, the mass of the electron could be calculated. ________

(f) The charge-to-mass ratio for positive particles is always larger than the charge-to-mass ratio for the electron. ________

(g) An alpha particle is the same as an electron. ________

(h) Gamma rays are not particles, but instead are high energy light waves. ________

(i) In the mass spectrometer the positive particle with the largest e/m ratio is the proton. ________

(j) The diameter of the nucleus of an atom is approximately 1/100,000 of the diameter of the atom. ________

(k) The number of protons in the nucleus of an atom is given by the mass number. ________

(l) Isotopes of a given element have the same number of protons but differ in the number of neutrons. ________

(m) Moseley found that he could assign an atomic number to an atom on the basis of the frequency of X-rays emitted by the atom. ________

2. The width of a sharp pencil point is about 0.1 mm. How many angstroms is this? The diameter of a carbon atom is about 1.54 Å. How many carbon atoms lay end-to-end across the pencil point?

_______________ _______

3. How many protons are there in these atoms?

(a) $^{32}_{16}S$ _______ (b) $^{192}_{77}Ir$ _______ (c) $^{39}_{19}K$ _______

4. How many neutrons are there in these atoms?

(a) $^{108}_{47}Ag$ _______ (b) $^{209}_{83}Bi$ _______ (c) $^{19}_{9}F$ _______

5. (Fill in the blanks)

(a) An atom that has acquired a charge by the gain or loss of electrons is called

____

(b) The relative charge on an α-particle is ____________________

(c) The relative charge on a γ-ray is ____________________

(d) The SI unit of frequency is called ____________________

(e) When a current of 1 amp flows for 1 sec the amount of charge that passes a given point in a wire is called

____

6. The wavelength of a yellow light wave is 589 nm (nanometer). What is its frequency?

____

7. The frequency of a green light wave is 5.49×10^{14} sec^{-1}. What is its wavelength in:

(a) centimeters ____________________

(b) nanometers ____________________

8. Antimony occurs in nature as a mixture of two isotopes, 57.25% ^{121}Sb with a mass of 120.904 amu and 42.75% ^{123}Sb with a mass of 122.904 amu. What is the average atomic mass of Sb?

____

3.10 THE PERIODIC LAW AND THE PERIODIC TABLE

Objectives

To understand the rationale behind the construction of the periodic table. You should learn the nomenclature that applies to the periodic table and the names applied to different sets of elements.

Review

The periodic table is without a doubt one of the most useful devices available to a chemist or a chemistry student. Trends in a large number of chemical and physical properties can be correlated with the positions of atoms within this table. Your ability to use the periodic table effectively depends on how well you understand its construction.

Remember that the elements are arranged in horizontal rows (called periods) in order of increasing atomic number. The periodic law states that when arranged in this manner the elements exhibit a periodic recurrence of properties. An important feature of the periodic table is that elements with similar chemical properties are arranged in vertical columns (called groups).

This section introduces you to a set of nomenclature associated with the periodic table. The important terms are given in boldface type in the text and are listed on the following page. You should learn the meanings of these terms.

Self-Test

9. Which of the following are representative elements: Cl, Fe, Cu, Na, Xe?

____

10. Which of these are transition elements: As, Hg, Ti, Ge, Sr?

____

11. Which of these is a halogen: S, Sn, Br, Na, Mg?

____

12. Which of these is a noble gas: O, Ne, Na, Ca, Zn?

____

13. Which of these is an alkali metal: Li, B, C, F, Xe?

____

14. Which of these is an alkaline earth metal: Zn, C, Cs, Ba, Kr?

____

15. Which of these are inner transition elements: Np, Ru, F, As, Pm?

____

16. Which of these is a metalloid: S, Ni, Ge, He, Mg? ____________

17. Which are metals: Sr, Si, Cr, Ce, U, P? ____________

18. Which are nonmetals: S, Ga, P, Pr, I, K? ____________

New Terms

group	alkali metals
period	alkaline earth metals
periodic law	halogens
representative elements	noble gases
transition elements	metal
inner transition elements	metalloid
lanthanides	nonmetals
actinides	ductility
rare earths	malleability

3.11 ATOMIC SPECTRA

Objectives

To understand that the light emitted by an atom is composed of a number of discrete wavelengths, rather than the entire spectrum.

Review

A white-hot object like the sun or an incandescent lamp emits light of all colors to give a continuous spectrum. Excited atoms emit only certain wavelengths (colors) and produce a line spectrum. It is possible to find an equation that allows the calculation of the wavelengths of the lines. An important aspect of this equation is that it involves the difference between the reciprocals of squares of integers. The occurrence of these integers provides the clue to the electronic structure of the atom.

Self-Test

19. What is the wavelength, in cm, of the second line of the Lyman series in the atomic spectrum of hydrogen. ____________

20. (Multiple choice) The existence of line spectra demonstrates that

(a) only certain electrons in atoms can be excited
(b) the electrons in an atom can have only certain specific energies
(c) Planck's equation doesn't always hold true
(d) white light is composed of many wavelengths
(e) none of these are correct ________

New Terms

continuous spectrum
line spectrum
Rydberg equation

3.12 THE BOHR THEORY OF THE HYDROGEN ATOM

Objectives

To show that the introduction of the idea of quantized energy levels in the atom permitted the explanation of atomic spectra. You should learn the relationship between frequency of light and energy.

Review

Planck had shown that the energy in a beam of light is proportional to the frequency of the light waves. Remember that $E = h\nu$.

The significance of Bohr's theory was that it introduced for the first time the idea that in an atom the electron is only permitted to have certain energies; intermediate energies are forbidden. Electrons change energy by going from one energy level to another. Energy levels can be identified by the value of a quantum number.

This section also illustrates how complex theories are tested. From the postulates of the theory an equation is derived, in this case an equation that can be used to calculate the wavelengths of lines in the atomic spectrum. This theoretical equation is compared to an equation based solely on the experimental data. If the equations match, it is taken as evidence for the validity of the theory; if they don't, the theory must be wrong. Bohr's success with hydrogen indicated he was on the right trail. The failure of his theory to predict the wavelengths of spectral lines of atoms more complicated than hydrogen demonstrated that there was a basic flaw somewhere in the theory.

Self-Test

21. (Multiple choice) Bohr's theory

    (a) proved that the electron travels in circular orbits about the nucleus

    (b) concluded that the radius of an orbit was inversely proportional to the quantum number, n

    (c) was successful in explaining the Rydberg equation for hydrogen

    (d) states that an electron gains energy when it moves from an orbit with a given value of n to an orbit with a smaller value of n

    (e) none of the above apply to Bohr's theory

____

New Terms

photon
energy level
quantum number

3.13 WAVE MECHANICS

Objectives

To understand that matter, like light waves, has wavelike properties. You should learn the meaning of wave function and orbital. Also learn the names and permissible values of the quantum numbers employed in wave mechanics.

Review

It has been shown experimentally that, as predicted by de Broglie, matter has wave properties. Proof lies in the diffraction of particles (Footnote 3 in the text).

Wave mechanics is the name of the theory that treats the electron as a wave. Solution of a wave equation gives a set of wave functions, ψ. Each wave function describes an atomic orbital which has a characteristic energy and which corresponds to a region around the nucleus where we are likely to find the electron. An orbital is identified by a set of three quantum numbers: n, l, and m. Review the values permitted for these quantum numbers on Page 72 of the text.

$$n = 1, 2, \ldots,$$

$$l = 0, 1, 2, \ldots, n-1$$

$$m = 0, \pm 1, \pm 2, \ldots, \pm l$$

Remember that subshells are identified by their value of l.

l	0	1	2	3
letter designation	s	p	d	f

Before moving on, examine the energy level diagram in Figure 3.20. Note that within a shell the energy of the subshells vary as: $s < p < d < f$. Also note that an s subshell consists of one orbital; a p subshell, three orbitals; a d subshell, 5 orbitals; and an f subshell, seven.

Self-Test

22. What values of m are allowed in the following?

    (a) 2s subshell ____________

    (b) 3d subshell ____________

    (c) 5f subshell ____________

    (d) 4p subshell ____________

23. Give the proper subshell designation corresponding to the following sets of quantum numbers.

    (a) n = 3, l = 1 ______

    (b) n = 4, l = 3 ______

    (c) n = 4, l = 2 ______

    (d) n = 2, l = 0 ______

24. Why don't we see wave properties for large particles like cars and baseballs? ____________________

New Terms

wave mechanics
quantum mechanics
wave function
orbital
principal quantum number
azimuthal quantum number
magnetic quantum number
subshell
ground state

3.14 ELECTRON SPIN AND THE PAULI EXCLUSION PRINCIPLE

Objectives

To see that the electron behaves as if it were spinning about its axis like a top. You should learn that the Pauli exclusion principle limits the number of electrons per orbital to two. You should learn how electron spin influences the magnetic properties of substances.

Review

The electron behaves like a tiny electromagnet, implying that it is spinning about its axis. There are two values of the spin quantum number s, +1/2 and -1/2, corresponding to two directions of rotation.

The Pauli exclusion principle requires that any two electrons in an atom have different sets of values for its four quantum numbers: n, l, m and s. If the first three are identical for two electrons, the electrons must spin in opposite directions.

The maximum number of electrons that can be placed in a given orbital is two, and they must have opposite spins. The maximum population per subshell is:

subshell	max population
s	2
p	6
d	10
f	14

It sometimes helps to remember this if you realize that the numbers 2, 6, 10, 14 form an arithmetic progression, each successive number being four larger than the one before it.

You should review the magnetic properties of substances as they are determined by the electron's spin.

Self-Test

25. (Multiple choice) One electron in an atom has the quantum numbers: n = 3, l = 2, m = -1, s = 1/2. Which of the following is not a possible set of quantum numbers for a second electron in this same atom?

(a) n = 1, l = 0, m = 0, s = -1/2
(b) n = 2, l = 1, m = -1, s = 1/2
(c) n = 3, l = 2, m = -1, s = 1/2
(d) n = 3, l = 1, m = -1, s = 1/2

(e) $n = 2, \; l = 0, \; m = 0, \; s = -1/2$ ________

26. (Fill in the blanks) The maximum number of electrons in

(a) the 2s subshell is ________

(b) the 3p subshell is ________

(c) the 6g subshell is ________

27. A neutral potassium atom must be paramagnetic. Why?

____

New Terms

Pauli exclusion principle	paramagnetic
diamagnetic	ferromagnetic

3.15 THE ELECTRON CONFIGURATIONS OF THE ELEMENTS

Objectives

To write electron configurations for the elements. You should learn both the conventional notation (e.g., $1s^2$...) as well as how to construct an orbital diagram.

Review

The number of electrons in a given subshell is specified by writing the subshell designation with the number of electrons indicated as an exponent. Thus $3p^4$ indicates four electrons in a 3p subshell.

Subshells in an atom become populated starting with the lowest energy level first. The sequence in which subshells become filled is determined by the energy level diagram in Figure 3.20 (Page 74 of the text).

When writing orbital diagrams, arrows are used to indicate electrons (head up for one direction of spin, and head down for the other). For example, the orbital diagram for boron (Z = 5) is

	B	↑↓	↑↓	↑	___	___
		1s	2s		2p	
or	B	[He]	↑↓	↑	___	___
			2s		2p	

Remember that [He] stands for the filled noble gas core. In a similar fashion we would write the orbital diagram for Ca as

$$\text{Ca} \quad [\text{Ar}] \quad \underset{4s}{\underline{\uparrow\downarrow}}$$

When more than one electron occupies a p, d or f subshell, Hund's rule applies, which tells us that the electrons are spread out over the orbitals as much as possible with their spins in the same direction. For example, the orbital diagram for phosphorus is

$$\text{P} \quad [\text{Ne}] \quad \underset{3s}{\underline{\uparrow\downarrow}} \quad \underbrace{\underline{\uparrow}\ \underline{\uparrow}\ \underline{\uparrow}}_{3p}$$

A phenomenon that has some important consequences in terms of chemical properties is that half-filled and filled subshells are extra stable. This causes Cr and Cu to have unexpected electron configurations.

New Terms

electron configuration	core electrons
orbital diagram	Hund's rule

3.16 THE PERIODIC TABLE AND ELECTRON CONFIGURATIONS

Objectives

To learn to use the periodic table to deduce the electron configuration of an element.

Review

The structure of the periodic table is a direct consequence of the order in which subshells are filled and, as described in the text, you can use the periodic table to help you write down electron configurations.

Example 3.2

Write the electron configuration of germanium (Z = 32).

Solution

There are 32 electrons in the atom. Moving from left to right across successive periods we fill:

$1s^2$ (period 1)
$2s^2$ (period 2 - Groups IA and IIA)
$2p^6$ (period 2 - Groups IIIA through the noble gases)
$3s^2$ (period 3 - Groups IA and IIA)
$3p^6$ (period 3 - Groups IIIA through the noble gases)
$4s^2$ (period 4 - Group IA and IIA)
$3d^{10}$ (period 4 - first row of transition elements
$4p^2$ (period 4 - Groups IIIA and IVA)

Writing all this together:

$$\text{Ge} \quad 1s^2 2s^2 2p^6 3s^2 3p^6 4s^2 3d^{10} 4p^2$$

Some people prefer to write all subshells of a given shell together.

$$\text{Ge} \quad 1s^2 2s^2 2p^6 3s^2 3p^6 3d^{10} 4s^2 4p^2$$

Self-Test (Sections 3.15 and 3.16)

28. Predict the electron configuration of

(a) Mg ____________________

(b) Cl ____________________

(c) Tc ____________________

(d) Ni ____________________

29. Construct orbital diagrams for the following:

(a) Si

(c) Fe

(b) Ca

(d) S

New Terms

none

3.17 THE SPATIAL DISTRIBUTION OF ELECTRONS

Objectives

To understand how wave mechanics describes the spatial distribution of electrons (i.e., where the electron is likely to be observed).

Review

Because of the Heisenberg uncertainty principle, wave mechanics describes the probability of finding the electron at points around the nucleus. The electron is viewed as being smeared out. Regions where the probability of finding the electron is high are said to have a high electron density.

The shapes of the probability distributions for different types of orbitals will be important in discussion of bonding in Chapter 16. Remember that s orbitals are spherical; p orbitals are dumbbell shaped. You should also know that the three orbitals in a p subshell are oriented at 90° to each other.

Notice that 2s and higher s orbitals contain nodes (where the electron density drops to zero), as do 3p and higher p orbitals. The important point, however, is that s orbitals have an overall spherical shape while the p orbitals tend to "point" in specific directions.

New Terms

uncertainty principle
probability distribution

3.18 THE VARIATION OF PROPERTIES WITH ATOMIC STRUCTURE

Objectives

To be able to predict trends in atomic size, ionization energy and electron affinity.

Review

Ionization energy is the energy needed to remove an electron from an isolated atom or ion.

Electron affinity is the energy released when an electron is added to a gaseous atom.

Within the periodic table atomic size generally decreases from left to right in a period and increases from top to bottom within a group. Variations in ionization energy and electron affinity generally parallel variations in size. Large atoms have low ionization energies and low electron affinities while small atoms have high ionization energies and high electron affinities. The variation of properties within the periodic table is summarized below.

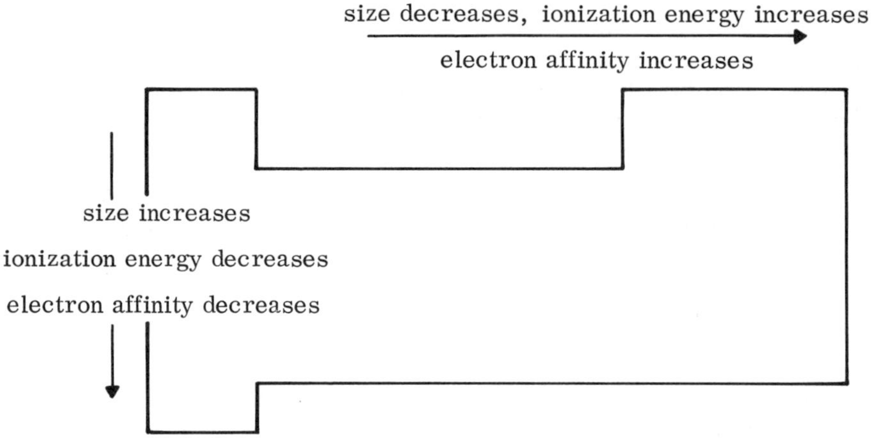

Remember that the second ionization energy is always larger than the first, the third larger than the second, and so forth. Also remember that the second electron affinity is always an endothermic quantity.

Remember that positive ions are always smaller than the neutral atoms from which they are formed and that an atom gets larger when it becomes a negative ion.

<u>Self-Test</u>

30. Which is the largest atom: C, N, Si, P? ________

31. Which atom has the largest ionization energy: B, C, Al, Si? ________

32. Which atom has the smallest electron affinity: B, C, Al, Si? ________

33. In each of the following, choose the larger.

    (a) Na or Na^+ ________

    (b) Cl or Cl^- ________

<u>New Terms</u>

effective nuclear charge	ionization energy
lanthanide contraction	electron affinity

ANSWERS TO SELF-TEST QUESTIONS

1. (a) False (b) True (c) False (d) True (e) True (f) False (g) False (h) True (i) True (j) True (k) False (l) True (m) True
2. 1×10^6 Å, 6.49×10^5 atoms C 3. (a) 16 (b) 77 (c) 19 4. (a) 61 (b) 126 (c) 10 5. (a) ion (b) +2 (c) zero (d) hertz (e) coulomb
6. 5.09×10^{14} Hz 7. (a) 5.46×10^{-5} cm (b) 546 nm 8. 121.8
9. Cl, Na, Xe 10. Hg, Ti 11. Br 12. Ne 13. Li 14. Ba 15. Np, Pm
16. Ge 17. Sr, Cr, Ce, U 18. S, P, I 19. 1.03×10^{-5} cm 20. b
21. c 22. (a) 0 (b) 0, +1, -1, +2, -2 (c) 0, +1, -1, +2, -2, +3, -3 (d) 0, +1, -1 23. (a) 3p (b) 4f (c) 4d (d) 2s 24. Their large masses give them very small wavelengths. 25. c 26. (a) 2 (b) 6 (d) 18
27. An odd number of electrons; they all can't be paired.
28. (a) Mg[Ne] $3s^2$ (b) Cl[Ne] $3s^2 3p^5$ (c) Tc[Kr] $5s^2 4d^5$ (d) Ni[Ar] $4s^2 3d^8$
29. (a) Si [Ne] ↑↓ (3s) ↑ ↑ ___ (3p) (b) Ca [Ar] ↑↓ (4s)
(c) Fe [Ar] ↓↑ (4s) ↓↑ ↑ ↑ ↑ ↑ (3d) (d) S [Ne] ↓↑ (3s) ↓↑ ↑ ↑ (3p)
30. Si 31. C 32. Al 33. (a) Na (b) Cl^-

4 CHEMICAL BONDING: GENERAL CONCEPTS

The discussion of chemical bonding in your text is divided between two chapters. This first chapter provides you with some general concepts and a simplified treatment of the subject. After you have had a chance to digest this material, some of the more modern bonding theories are described in Chapter 17.

Chemical bonds are what hold atoms together in chemical compounds. In this chapter you should learn the types of chemical bonding, how they relate to electronic structure, and how they influence the properties of compounds. The weaker attractive forces that exist between neighboring molecules are also discussed. It is these attractive forces that determine the physical properties of substances.

4.1 LEWIS SYMBOLS

Objectives

To learn how to write the Lewis symbol, or dot symbol, for atoms of the representative elements.

Review

Remember that it is the outer shell (valence shell) electrons that are involved in the formation of chemical bonds. Lewis symbols are simply a bookkeeping device that are used to keep track of the valence electrons during bond formation. Dots (or some other symbol such as an x or a circle) are used to represent valence electrons. Review Table 4.1 in the text before working on the following Self-Test.

Self-Test

1. Write Lewis symbols for:

   (a) Si __________ (c) Sr __________

   (b) Cl __________ (d) As __________

New Terms

valence shell
Lewis symbol
electron-dot formula

4.2 THE IONIC BOND

Objectives

To understand what an ionic bond is, how it is formed, and the types of elements that form ionic bonds.

Review

Remember that the ionic bond is formed by electron transfer. This produces ions that attract each other because of their opposite charges. Among the representative elements atoms tend to gain or lose electrons until they achieve an electron configuration that is identical to a noble gas. A noble gas atom has eight electrons in its outer shell, hence the octet rule which states that atoms tend to gain or lose electrons until there are eight electrons in the outer shell. You should be able to diagram the formation of an ionic compound using Lewis symbols. The Self-Test at the end of this section provides some practice.

Among the transition elements the octet rule doesn't always hold - other relatively stable electron configurations also exist. Review Table 4.2 in your text.

Ionic compounds exist between polyatomic ions too. Learn the formulas and names of the ions in Table 4.3. When writing the formulas and compounds containing these ions remember that the ratio of positive to negative ions must be chosen to give a neutral formula. For instance, the compound containing Fe^{3+} and CO_3^{2-} has the formula, $Fe_2(CO_3)_3$. A simple way to get this formula is to use the number of charges on one ion as the subscript on the other.

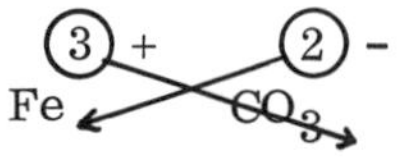

gives $Fe_2(CO_3)_3$

This system works for writing the formula for any ionic compound. Just remember to reduce the subscripts to the smallest set of whole numbers. For example,

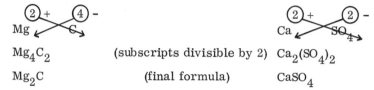

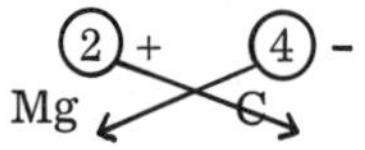

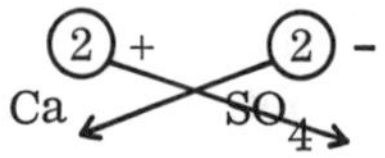

Mg_4C_2	(subscripts divisible by 2)	$Ca_2(SO_4)_2$
Mg_2C	(final formula)	$CaSO_4$

Self-Test

2. Use Lewis symbols to diagram the formation of an ionic bond between:

   (a) K and Cl ____________________

   (b) Na and O ____________________

3. Write the pseudonoble gas configuration for Zn^{2+}.

   ____

4. Write the formulas for the ionic compounds formed from the ions:

   (a) Cr^{2+} and PO_4^{3-} ________________

   (b) NH_4^+ and SO_4^{2-} ________________

   (c) Ga^{3+} and ClO_2^- ________________

   (d) Cr^{3+} and $C_2O_4^{2-}$ ________________

   (e) NH_4^+ and S^{2-} ________________

New Terms

ionic bond
electrovalent bond
cation
anion
octet rule
pseudonoble gas configuration
polyatomic ion

4.3 FACTORS INFLUENCING THE FORMATION OF IONIC COMPOUNDS

Objectives

To examine the factors that favor ionic bonding and to learn what gives rise to the stability of ionic compounds.

Review

An ionic compound is stable because of the very large lattice energy that is released when the ions come together to produce the ionic solid. If it were not for this lattice energy, ionic compounds would not exist. Thus, in the gas phase the ions, $Li^+(g)$ and $F^-(g)$, are less stable (i.e., of higher energy) than Li(g) and F(g) and therefore Li(g) and F(g) would not react spontaneously to produce the ions.

Ionic compounds tend to form most readily when metals of low ionization energy and electron affinity react with nonmetals of high ionization energy and electron affinity, and when the lattice energy of the resultant compound is high. Therefore, the metals on the extreme left of the periodic table (Groups IA and IIA) and the nonmetals on the extreme upper right tend to form ionic compounds.

New Terms

lattice energy
Born-Haber cycle

4.4 THE COVALENT BOND

Objectives

To learn how the covalent bond is formed. You should learn how to draw Lewis structures for covalent molecules. This should include molecules and polyatomic ions that contain single, double and triple bonds.

Review

A single covalent bond consists of a pair of electrons shared between two atoms. This is favored when the difference between ionization energies and electron affinities of the combining atoms is not large. By electron sharing atoms usually complete their octet. The octet rule is used when we draw Lewis structures for molecules. Try to become adept at this; you'll find it very helpful later in the course.

Let's look at some rules that can help you draw Lewis structures. First, you must know what atoms are bonded together. Usually there is some central atom bonded to a number of others and in the formula of many compounds the central atom is written first. For example, in PCl_3 the P atom is bonded to three Cl atoms and each Cl is bonded only to the central P atom.

Once you've established the arrangement of atoms, follow the procedure below:

(1) Count up all of the valence electrons for the atoms in the molecule or ion. If you are working with an ion, add one electron for each negative charge on the ion; deduct one electron for each positive charge. For example,

NO_2^+	has	$5e^-$	(for N atom)
		$+12e^-$	(for two O atoms)
		$-\ 1e^-$	(for + charge)
	Total	16	valence electrons

(2) Place a pair of electrons in each bond. Again, for NO_2^+,

O — N — O

This leaves 16 - 4 = 12 electrons to go.

(3) Fill the octets of the outer atoms first with the remaining electrons. If electrons are left, add them to the central atom. For NO_2^+,

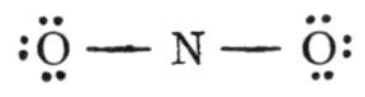

Notice that here we have run out of dots before we could fill the octet of the nitrogen.

(4) If you run out of electrons before you fill the octet of the central atom, you must create multiple bonds. For example,

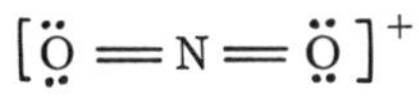

(5) If you have electrons left over after you've filled the octets of all of the atoms, place the extra electrons on the central atom. It is the central atom that will violate the octet rule. For example, with SF_4 we have 6 + 28 = 34 valence electrons. Applying these rules we generate the following:

(lay out the F's around the S)

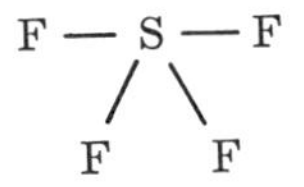

(1 pair in each bond. This leaves 34 - 8 = 26 dots to go.)

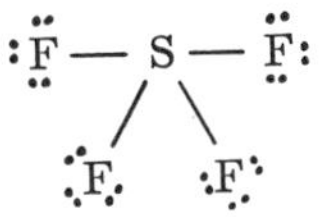

(fill the octets of the F's. All atoms have an octet, but we still have 2 electrons left over.)

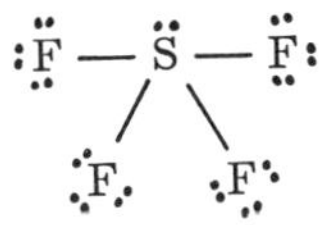

(place the extra 2 electrons on the S. The sulfur violates the octet rule. This is the dot structure of SF_4.)

Self-Test

5. Using the Lewis symbols for the elements Si, P, S and Cl predict the formulas for the simplest compounds these elements would form with hydrogen and write their Lewis structures.

6. Write electron-dot formulas for the following (the central atom is written first in each formula).

(a) $SiCl_4$

(b) ClF_3

(c) SCl_2

(d) SO_4^{2-}

(e) CO

(f) ClO_2^-

New Terms

covalent bond
single bond
double bond
triple bond

4.5 RESONANCE

Objectives

To learn the meaning of the term resonance and to learn to draw resonance structures where applicable.

Review

Remember that the actual structure of a resonance hybrid never corresponds to any of the resonance forms that you draw for the molecule or ion; it is always something in between. Resonance structures arise when there is more than one reasonable way of distributing electron pairs in a molecule. The NO_3^- ion, for example, is a resonance hybrid of three contributing structures:

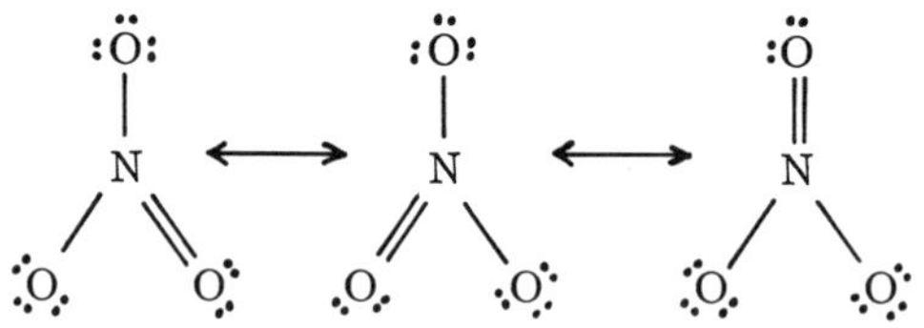

When you draw the dot structure for a molecule or ion and find that you must create a double bond to complete the octet of every atom, and when there is a choice of where to form the double bond, resonance structures occur. For instance, in the dot structure for the NO_3^- ion above, the double bond could be placed in any one of three places when you construct the dot formula. Therefore, three resonance structures occur. The actual structure of NO_3^- is a sort of average of these three. Each bond is approximately 1-1/3 bonds.

Self-Test

7. Draw all resonance structures for

   (a) SO_3

(b) N_3^- (structure, N·····N·····N)

(c) HCO_2^- (structure, H·····C(·····O)(·····O))

<u>New Terms</u>

resonance
resonance hybrid

4.6 COORDINATE COVALENT BONDS

<u>Objectives</u>

To see how atoms can use unshared electron pairs to form additional bonds.

<u>Review</u>

The coordinate covalent bond is a bookkeeping device that is sometimes convenient to use in accounting for the bonding in some compounds. Remember that the properties of a covalent bond do not depend on the origin of the electrons shared between the atoms.

<u>Self-Test</u>

8. Draw electron-dot formulas for the following showing how the structure can be explained in terms of coordinate covalent bonding.

(a) H_3O^+ (O is central atom)

(b) Cl_3PO (P is central atom)

(c) BF_4^- (B is central atom)

New Terms

coordinate covalent bond
dative bond
addition compound

4.7 BOND ORDER AND SOME BOND PROPERTIES

Objectives

To learn how three bond properties, bond length, bond strength (bond energy), and vibrational frequency, are related to the electron density between two atoms.

Review

Bond order is the number of electron pairs shared between two atoms. Remember that as the bond order increases,

(a) bond length decreases
(b) bond energy, which in a sense is the bond strength, increases
(c) vibration frequency increases

Self-Test

9. The bond length in carbon monoxide, CO, is 1.13 Å. Judging from the data in Table 4.4 of your text, what can you say about the bond order in CO?

____

10. Arrange the following in order of predicted decreasing C-O bond length. Specify the average C-O bond order in each.

(a) :Ö═C═Ö:

(b) :C≡O:

(c) 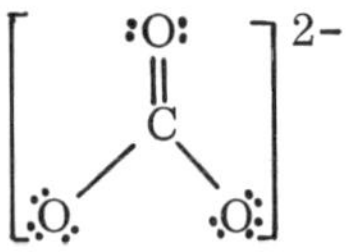 (one of three resonance structures)

(d) 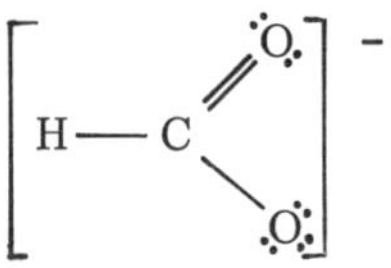 (one of two resonance structures)

(e)

$$
\begin{array}{c}
\quad\;\; H \\
\quad\;\; | \\
H - C - \overset{\displaystyle..}{\underset{\displaystyle..}{O}} - H \\
\quad\;\; | \\
\quad\;\; H
\end{array}
$$

_____

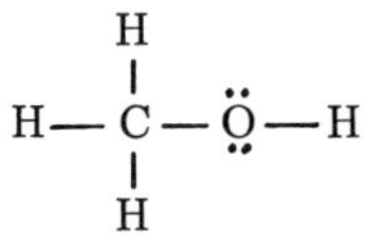

11.   What kind of information is obtained from the infrared absorption
spectrum of a molecule? _____

New Terms

bond order   bond energy
bond length   vibrational frequency

4.8   POLAR MOLECULES AND ELECTRONEGATIVITY

Objectives

To understand that in most molecules electrons are not shared
equally between atoms because different atoms have different
tendencies to attract electrons.

Review

Remember that electronegativity refers to the attraction an atom
has for electrons in a bond and that it is the difference in electronegativity
that determines the polarity of a bond, as well as which end of a polar bond
carries the negative charge. An important point to remember from this
section is that bonds can vary anywhere between essentially 100% covalent
to essentially 100% ionic.

You should also know the molecular shapes that produce nonpolar
molecules, even though the molecules may have polar bonds. In general,
if you draw the electron-dot formula for a molecule with the general formula AX_n (all X's identical), and all of the electrons around the central atom
are used for bonding, the molecule will be nonpolar.

Self-Test

12.   Use Table 4.6 in the text to determine which end of the following bonds
carries a partial negative charge.

(a) Sb-H _____   (c) C-O   _____

(b) P-S   _____   (d) Br-S _____

13. Which of the following molecules are nonpolar?

(a) NH_3 ________ (d) PCl_3 ________

(b) TeF_6 ________ (e) SO_3 ________

(c) $SiCl_4$ ________

New Terms

electronegativity
dipole
dipole moment
polar

4.9 OXIDATION AND REDUCTION, OXIDATION NUMBERS

Objectives

To learn how many chemical reactions can be explained in terms of the transfer of electrons, either partially or completely, from one atom to another. You should become familiar with, and be able to use oxidation numbers to keep tabs on electrons.

Review

Be sure you know the meaning of the terms oxidation and reduction. You should be able to identify the oxidizing agent and reducing agent in a chemical reaction. The definitions of oxidizing agent and reducing agent are sometimes confusing. Just remember that if a substance is oxidized (i.e., loses electrons), we call it a reducing agent; a substance that is reduced is an oxidizing agent.

It is important to learn the rules for assigning oxidation numbers. Practice on the examples in the Self-Test below.

Self-Test

14. Assign oxidation numbers to the atoms in the following formulas.

(a) $KClO_3$ ____________________

(b) MnO_4^{2-} ____________________

(c) $S_2O_3^{2-}$ ____________________

(d) $SiCl_4$ ____________________

(e) BrF_3 ____________________

(f) P_4O_6 ____________________

(g) $SbCl_6^-$ ____________________

(h) $C_{12}H_{22}O_{11}$ ____________________

(i) S_3^{2-} ____________________

(j) O_3 ____________________

15. Identify the oxidizing and reducing agent in the following reactions.

(a) $H_2 + Cl_2 \longrightarrow 2HCl$ ____________

(b) $2Na_2S_2O_3 + I_2 \longrightarrow 2NaI + Na_2S_4O_6$ ____________

(c) $3\,O_2 + C_2H_4 \longrightarrow 2CO_2 + 2H_2O$ ____________

(d) $K_2Cr_2O_7 + 14HCl \longrightarrow 3Cl_2 + 2KCl + 2CrCl_3 + 7H_2O$

____

New Terms

oxidation
reduction
redox
oxidizing agent
reducing agent
oxidation number
oxidation state

4.10 THE NAMING OF CHEMICAL COMPOUNDS

Objectives

To be able to name simple inorganic compounds.

Review

The rules for naming inorganic compounds are given in Appendix B of the text. If you have had chemistry in high school, you should remember many of them. If you've never had chemistry before, it is important that you learn how to name simple inorganic compounds. Besides increasing your ability to follow what's going on in class, you will probably need to know the names of chemicals if you ever have to get them from the storeroom or order them from a catalog.

Self-Test

16. Name the following:

(a) $NiCl_2$ ____________________

(b) $Cr_2(CO_3)_3$ ____________________

(c) $Sr(NO_3)_2$ ____________________

(d) K_2SO_4 ____________________

(e) SF_6 ____________________

(f) CCl_4 ____________________

17. Write formulas for the following:

(a) tin(IV) oxide __________

(b) boron trichloride __________

(c) calcium bicarbonate __________

(d) iron(III) oxide __________

(e) dinitrogen pentoxide __________

(f) sodium phosphide __________

New Terms

trivial name

4.11 OTHER BINDING FORCES

Objectives

To examine the weaker binding forces that exist between molecules, rather than the strong binding forces that exist within molecules.

Review

The chemical bonds discussed in earlier sections are responsible for the chemical properties of substances. The intermolecular attractions (attractions between neighboring molecules) are responsible for the physical properties of compounds.

Dipole attractions exist between polar molecules. An extra strong type of dipole attraction called hydrogen bonding occurs when hydrogen is bonded to a very small, very electronegative element (N, O, F). London forces are generally weaker than dipole attractions and exist between all molecules and ions. Their presence is usually noticed most when other types of attractive forces are absent.

Keep in mind the effect that hydrogen bonding has on the structure of water, particularly how it gives ice an "open" structure that is less dense than liquid water.

Self-Test

18. Predict what kinds of intermolecular attractive forces would be observed in the following substances:

(a) PCl_3 _____

(b) CO_2 _____

(c) HF _____

New Terms

dipole attractions
hydrogen bonding
London forces

ANSWERS TO SELF-TEST QUESTIONS

1. (a) ·Ṡi· (b) :C̈l· (c) ·Sr· (d) :Äs·

2. (a) K× + ·C̈l: ⟶ K⁺ , $\left[\overset{×}{\underset{..}{C}l:}\right]^{-}$ (b) Na× + ·Ö: ⟶ 2Na⁺, $\left[\overset{×}{\underset{..}{O}:}\right]^{2-}$
   Na×

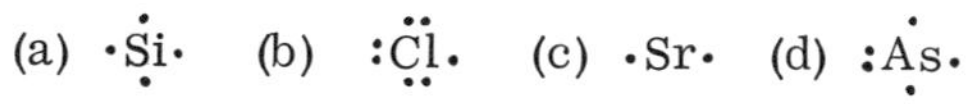

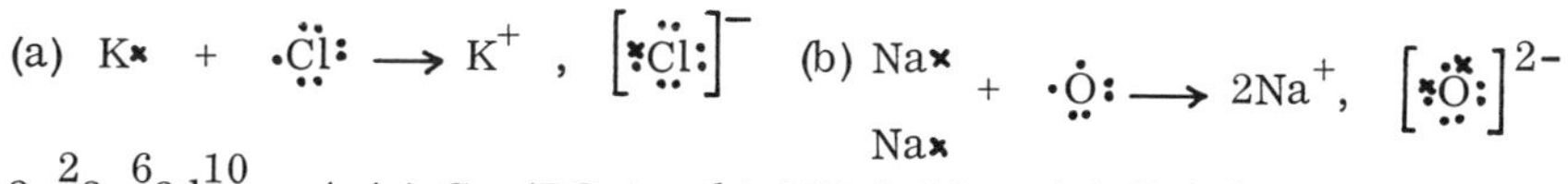

3. $3s^2 3p^6 3d^{10}$ 4. (a) $Cr_3(PO_4)_2$ (b) $(NH_4)_2SO_4$ (c) $Ga(ClO_2)_3$
   (d) $Cr_2(C_2O_4)_3$ (e) $(NH_4)_2S$

5. SiH_4,   H   , PH_3, H : P̈ : H , H_2S, H : S̈ : H,
           H : Si : H   H
              H

HCl, :C̈l : H 6. (a)   :C̈l: (b) :C̈l—F̈:   (c) :C̈l—S̈—C̈l:
                         :C̈l— Si —C̈l: F̈. .F̈.
                             :C̈l:

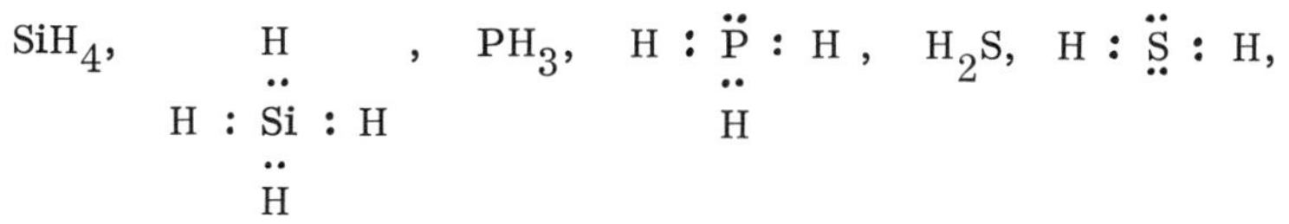

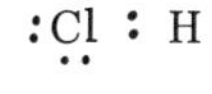

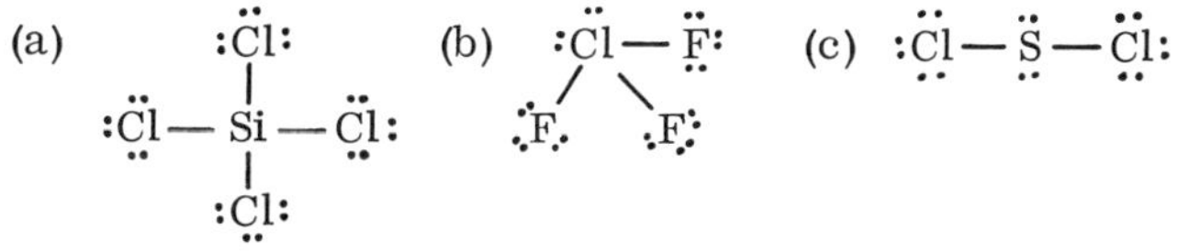

(d) $\left[\begin{array}{c} :\!\ddot{O}\!: \\ | \\ :\!\ddot{O}\!-\!S\!-\!\ddot{O}\!: \\ | \\ :\!\ddot{O}\!: \end{array}\right]^{2-}$ (e) $:\!C\!\equiv\!O\!:$ (f) $\left[:\!\ddot{O}\!-\!\ddot{C}l\!-\!\ddot{O}\!:\right]^{-}$

7. (a)

(b) $:\!N\!\equiv\!N\!-\!\ddot{N}\!:\ \leftrightarrow\ :\!\ddot{N}\!-\!N\!\equiv\!N\!:\ \leftrightarrow\ :\!\ddot{N}\!=\!N\!=\!\ddot{N}\!:$

(c)

8. (a) $H \overset{\times}{\underset{\times}{:}} \overset{\cdot\cdot}{O} : H$ (b)

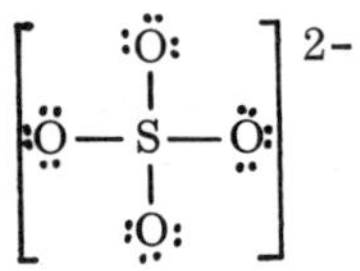

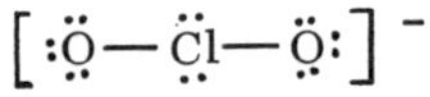

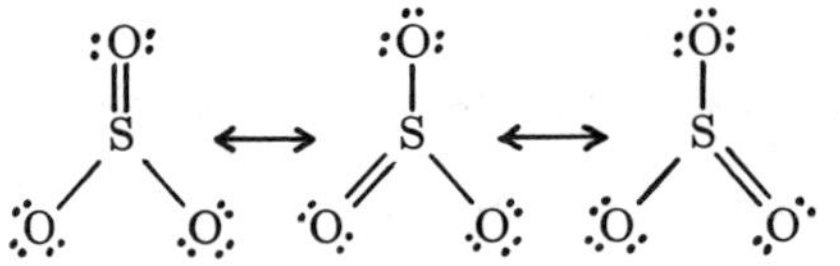

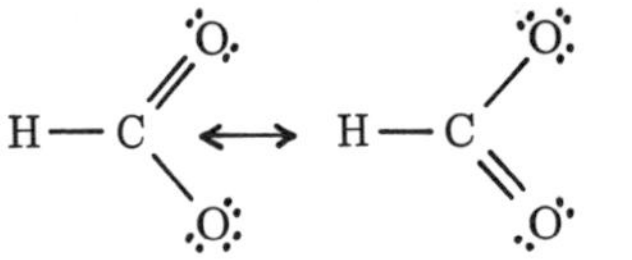

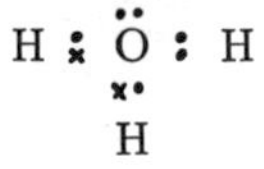

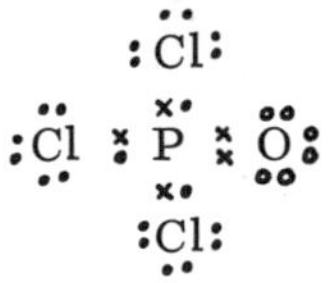

(c)

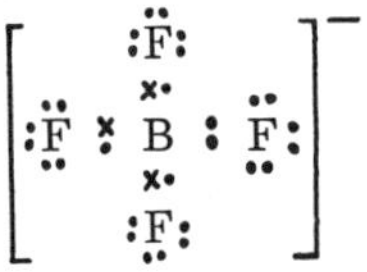

Notice that to identify coordinate covalent bonds it
is necessary to keep track of the <u>origin</u> of all of the
valence electrons in the molecule.

9. It must be approximately 3.   10. e > c > d > a > b   11. The vibration
frequencies of the bonds   12. (a) H (b) S (c) O (d) Br   13. TeF_6, $SiCl_4$,
SO_3 14. (a) K, +1; Cl, +5; O, -2 (b) Mn, +6; O, -2 (c) S, +2; O, -2
(d) Si, +4; Cl, -1 (e) Br, +3; F, -1 (f) P, +3; O, -2 (g) Sb, +5; Cl, -1
(h) C, 0; H, +1; O, -2 (i) S, -2/3 (j) 0   15. Oxidizing Agent (a) Cl_2 (b) I_2
(c) O_2 (d) $K_2Cr_2O_7$ Reducing Agent (a) H_2 (b) $Na_2S_2O_3$ (c) C_2H_4
(d) HCl   16. (a) nickel(II) chloride (b) chromium(III) carbonate
(c) strontium nitrate (d) potassium sulfate (e) sulfur(IV) fluoride or sulfur
hexafluoride (preferred)   (f) carbon(IV) chloride or carbon tetrachloride
(preferred)   17. (a) SnO_2 (b) BCl_3 (c) $Ca(HCO_3)_2$ (d) Fe_2O_3 (e) N_2O_5
(f) Na_3P
18. (a) dipole–dipole + London (PCl_3 is polar)
    (b) London (CO_2 is nonpolar)
    (c) hydrogen bonding + London (H is bonded to very electronegative F)
Notice that London forces are always present.

5 CHEMICAL REACTIONS IN AQUEOUS SOLUTION

Most chemical reactions are carried out in solution and very often the solvent is water. This chapter focuses on a variety of aspects dealing with reactions in water solutions. The chemistry described here is very important and must be learned thoroughly. Once again, if you have had high school chemistry, you may already know much of the material in this chapter. Review it anyway and test yourself to be sure. If you've not had high school chemistry, be sure to master this material well. You should be able to answer the self-test questions without having to look at rules or tables, unless directed to do so in the question.

5.1 SOLUTION TERMINOLOGY

Objectives

To become familiar with terms commonly used to describe solutions and their components.

Review

Be sure you understand and are familiar with the terms introduced in this section so that you can follow discussions when they are used later on.

Self-Test

1. Sugar has a solubility of 211 g per 100 g of water at 25°C. A solution containing 215 g of sugar in 100 g of water at 25°C would be described

by which of the following terms?
(a) concentrated (b) dilute (c) unsaturated
(d) supersaturated (e) saturated ________

New Terms

solvent
solute
concentrated
dilute
solubility
saturated
unsaturated
supersaturated

5.2 ELECTROLYTES

Objectives

To understand what takes place when an ionic solid dissolves in water. You will see that these substances break apart in water, as do certain molecular substances, to produce ions that are free to roam about.

Review

Remember that essentially all ionic solids are virtually 100% dissociated in aqueous solution. The ions are surrounded by water molecules and are said to be hydrated. Some covalent substances also form ions in aqueous solution by reaction with the solvent. An important ion produced in this way is the hydronium ion, H_3O^+. If dissociation is nearly complete, substances are said to be strong electrolytes; if only partially dissociated, they are said to be weak electrolytes. Undissociated covalent substances are nonelectrolytes.

New Terms

dissociation
electrolyte
hydronium ion
hydrated ion
strong electrolyte
weak electrolyte
nonelectrolyte

5.3 CHEMICAL EQUILIBRIUM

Objectives

To learn that chemical reactions are able to proceed in two directions, from reactants to products and from products to reactants. You should become familiar with the concept of dynamic equilibrium.

Review

The concept of dynamic equilibrium in chemistry is extremely important. Remember that chemical reactions are generally able to proceed in both forward and reverse directions. When opposing reactions are occurring at the same speed there is no change in the amounts of reactants or products. We use the term, position of equilibrium, to describe how far toward completion the reaction proceeds before equilibrium is reached. In a reaction

$$A \rightleftharpoons B$$

the position of equilibrium lies far to the right if a lot of B is produced from A by the time equilibrium is attained.

Self-Test

2. Write chemical equations showing the equilibria in the dissociation of the following weak electrolytes (see Table 5.1):

   (a) NH_3 ______________________________

   (b) HCN ______________________________

New Terms

equilibrium
position of equilibrium

5.4 IONIC REACTIONS

Objectives

To describe ionic reactions in solution by chemical equations that depict reactions between ions. You should also learn how to predict the products of an ionic reaction on the basis of solubility or the possible formation of an undissociated product or a gas. You should learn the solubility rules given here.

Review

Remember that the ionic equation for a reaction is obtained by writing the formulas of all soluble ionic compounds in dissociated form. Insoluble compounds are written in undissociated form (the list of solubility rules in this section should be memorized so that it should not be necessary to turn to them constantly). An example is the reaction of sodium sulfate with barium nitrate.

$$Na_2SO_4 + Ba(NO_3)_2 \longrightarrow 2NaNO_3 + BaSO_4$$

This type of "exchange of partners" reaction is called metathesis. It is also often called a double replacement reaction. In this example three of the compounds are soluble (Na_2SO_4, $Ba(NO_3)_2$ and $NaNO_3$) while only one is insoluble ($BaSO_4$). Writing the soluble ones in dissociated form gives the ionic equation,

$$2Na^+ + SO_4^{2-} + Ba^{2+} + 2NO_3^- \longrightarrow 2Na^+ + 2NO_3^- + BaSO_4(s)$$

The net ionic equation is obtained by eliminating ions that appear the same on both sides of the arrow.

$$Ba^{2+} + SO_4^{2-} \longrightarrow BaSO_4(s)$$

An ionic reaction will proceed to completion if a net ionic equation can be written without all of the ions cancelling. For example, nothing is observed to happen when solutions of $CaCl_2$ and $Ba(NO_3)_2$ are mixed. The "reaction",

$$CaCl_2 + Ba(NO_3)_2 \longrightarrow BaCl_2 + Ca(NO_3)_2$$

does not occur since all reactants and products are soluble.

$$Ca^{2+} + 2Cl^- + Ba^{2+} + 2NO_3^- \longrightarrow Ba^{2+} + 2Cl^- + Ca^{2+} + 2NO_3^-$$

Notice that all ions cancel. Nothing is left, hence no reaction.

Remember that all ions won't cancel if:

(a) a product is insoluble.
(b) a product is a weak electrolyte.
(c) a product is a gas.

Substances that fit the last two conditions are in Tables 5.1, 5.2 and 5.3 of the text.

Sometimes a reactant will also be a solid or weak electrolyte. A reaction will occur in these instances only if a product is less soluble, a weaker electrolyte, or a gas. You are not expected to predict the outcome of these reactions now, since you haven't been given rules about relative solubilities or degrees of dissociation.

Self-Test

3. Study the solubility rules before answering this question. Use this question to test your knowledge of the rules. If the substance is soluble, write S; if it is insoluble, write I.

(a) KNO_3 ________ (e) ZnO ________

(b) $MgSO_4$ ________ (f) $PbSO_4$ ________

(c) AgI ________ (g) $Ni(OH)_2$ ________

(d) $FeCO_3$ ________ (h) $(NH_4)_2CrO_4$ ________

4. Write ionic and net ionic equations for these reactions:

(a) $NaI + AgNO_3 \longrightarrow AgI + NaNO_3$

(b) $Pb(NO_3)_2 + Ba(OH)_2 \longrightarrow Pb(OH)_2 + Ba(NO_3)_2$

(c) $AgCl + NaBr \longrightarrow AgBr + NaCl$

(d) $ZnCl_2 + Na_2CO_3 \longrightarrow 2NaCl + ZnCO_3$

(e) $BaCl_2 + 2NaBr \longrightarrow BaBr_2 + 2NaCl$

5. Write ionic and net ionic equations for these reactions:

(a) $CaCO_3 + 2HCl \longrightarrow H_2O + CO_2 + CaCl_2$

(b) $(NH_4)_2SO_4 + 2NaOH \longrightarrow 2NH_3 + 2H_2O + Na_2SO_4$

(c) $Na_2C_2O_4 + 2HCl \longrightarrow 2NaCl + H_2C_2O_4$

(d) $BaCO_3 + H_2SO_4 \longrightarrow BaSO_4 + H_2O + CO_2$

(e) $K_2SO_3 + H_2SO_4 \longrightarrow K_2SO_4 + H_2O + SO_2$

6. Write molecular, ionic, and net ionic equations for the reaction, if any, that would occur between the following:

(a) $NiCl_2$ and Na_2CO_3 (d) Na_2SO_4 and $Mg(NO_3)_2$

(b) $NaCl$ and $CaBr_2$ (e) Na_2S and HCl

(c) $MgCl_2$ and $NaOH$

New Terms

precipitate
metathesis
double replacement
molecular equation
ionic equation
net ionic equation
spectator ion

5.5 ACIDS AND BASES IN AQUEOUS SOLUTION

Objectives

To learn how acids and bases are defined and how acids and bases react with each other.

Review

Remember:

acids produce H_3O^+ when dissolved in water
bases produce OH^- when dissolved in water

Many acids contain hydrogen that attaches itself as H^+ to water molecules to produce H_3O^+. In many cases H_3O^+ is simply indicated as H^+ since this is the "active ingredient" in the H_3O^+ ion during chemical reactions. Metal hydroxides, when soluble, dissociate to give OH^- in solution.

Acids and bases react with each other to produce a salt and water. A salt is a general term that is used to refer to any ionic compound except metal hydroxides, which are called bases. An example of an acid-base neutralization reaction is

$$2HCl + Ba(OH)_2 \longrightarrow BaCl_2 + 2H_2O$$

Polyprotic acids (those able to supply more than one H^+ per molecule of acid) can be partially neutralized to give acid salts; for example, $NaHSO_4$ from the partial neutralization of H_2SO_4. Bases generally do not undergo partial neutralizations; instead, all of the OH^- of a base is neutralized in the formation of salts which can be isolated from solution.

You should also remember that metal oxides are basic. When they dissolve in water they produce hydroxides because of the reaction,

$$O^{2-} + H_2O \longrightarrow 2\ OH^-$$

Even if they are insoluble, metal oxides still react with acids; for example,

$$\underset{\text{(rust)}}{Fe_2O_3} + 6H^+ \longrightarrow 2Fe^{3+} + 3H_2O$$

Nonmetal oxides are acidic. Many give acids when they dissolve. For example,

$$CO_2 + H_2O \longrightarrow H_2CO_3$$

Self-Test

7. Write balanced molecular equations for the neutralization reaction between:
   (a) $Ca(OH)_2$ and HCl
   (b) NH_3 and H_2SO_4 (complete neutralization)
   (c) $Al(OH)_3$ and H_2SO_4 (complete neutralization)
   (d) NaOH and HNO_3
   (e) MgO and H_2SO_4
8. Write formulas for all salts formed between $Ca(OH)_2$ and
   (a) $H_2C_2O_4$ ____________________
   (b) H_2CO_3 ____________________
   (c) H_3AsO_4 ____________________
   (d) HBr ____________________

New Terms

acid
base
neutralization
monoprotic acid
polyprotic acid
acid salt

5.6 THE PREPARATION OF INORGANIC SALTS BY METATHESIS REACTIONS

Objectives

To learn methods that can be used to prepare salts by reactions in aqueous solution.

Review

The techniques discussed in this section call upon what you learned in Sections 5.4 and 5.5. Your ability to choose a method to prepare a salt obviously depends on how well you've learned this previous material. If you have difficulty with most of the problems in the Self-Test, go back and review Sections 5.4 and 5.5. In particular, pay attention to solubility rules and the factors that cause reactions to go to completion.

Precipitation Reactions. These make use of the formation of a precipitate to obtain the desired product. You need to remember the solubility rules to apply this method. Remember to use soluble reactants that give only one insoluble product.

Example 5.1

How can we prepare $CuCO_3$ by a precipitation reaction?

Solution

The product, $CuCO_3$, is insoluble; therefore we want the other product to be soluble. Also, we want to begin with a soluble carbonate and a soluble copper salt. Possible reactants are $CuCl_2$ and Na_2CO_3. Both are soluble and the choice of the sodium salt to provide the CO_3^{2-} ensures us that the other product in the metathesis reaction will be soluble.

$$\underset{\text{(sol.)}}{CuCl_2} + \underset{\text{(sol.)}}{Na_2CO_3} \longrightarrow \underset{\text{(sol.)}}{2NaCl} + \underset{\text{(insol.)}}{CuCO_3}$$

Neutralization Reactions. The desired salt is derived from the cation of a base and the anion of an acid. For instance, to prepare $CuBr_2$ you could use the reaction between $Cu(OH)_2$ and HBr,

$$Cu(OH)_2(s) + 2HBr(aq) \longrightarrow CuBr_2(aq) + 2H_2O$$

Since $Cu(OH)_2$ is insoluble, an excess of $Cu(OH)_2$ is used so that all of the HBr in solution is used up. Excess insoluble $Cu(OH)_2$ is removed by filtration and the solution that passes through the filter contains only $CuBr_2$ which can be recovered by evaporation.

This method is good because most metal hydroxides are insoluble and can be prepared from other readily available salts by reaction with a base. For instance,

$$Cu(NO_3)_2(aq) + 2NaOH(aq) \longrightarrow Cu(OH)_2(s) + 2NaNO_3(aq)$$

Reactions in Which a Product is a Gas. Reactions of metal carbonates were discussed in the text. Any metathesis reaction in which one product is a gas will leave the other product by itself in solution. Some sample reactions are:

$$K_2SO_3 + 2HClO_4 \longrightarrow 2KClO_4 + H_2O + SO_3(g)$$

$$FeS + 2HBr \longrightarrow FeBr_2 + H_2S(g)$$

Review also Examples 5.5 to 5.8 in the text before beginning the Self-Test.

Self-Test

9. How would you prepare the following by a precipitation reaction?

   (a) $Fe(NO_3)_3$ (b) AgBr (c) $BaBr_2$ (d) NaOH

10. How would you prepare the following by a neutralization reaction?

   (a) $Cu(HSO_4)_2$ from $CuCl_2$ (c) $Mg(NO_3)_2$ from $MgCl_2$

   (b) $NiSO_4$ from $NiCl_2$ (d) $Ca(NO_3)_2$ from CaO

11. How would you prepare the following by a reaction that produces a gas?

   (a) $FeCl_2$ from $FeCO_3$ (d) $ZnCl_2$ from ZnS

   (b) $Co(NO_3)_2$ from $CoCO_3$ (e) $Ca(NO_3)_2$ from $CaCl_2$

   (c) $NaNO_2$ from NH_4NO_2

New Terms

none

5.7 OXIDATION-REDUCTION REACTIONS

Objectives

To learn to balance oxidation-reduction reactions by making use of the changes of oxidation numbers.

Review

This section and the next discuss alternative ways of balancing redox reactions. The approach taken by the oxidation-number-change (ONC) method and the ion-electron (IE) method (Section 5.8) are different in several respects. It is only necessary to assign oxidation numbers in the ONC method. The IE method does not employ oxidation numbers even though the same end result is achieved. The key to both methods is making the number of electrons gained equal to the number lost.

In the ONC method, be sure to calculate the number of electrons transferred per formula unit for the reactants. Place coefficients in the equation to make the total electron loss equal to the total electron gain. Balance the remainder of the equation by inspection.

Remember, there is never any reason to be unsure that an equation is balanced correctly. You can always count up the numbers of each kind of atom on each side of the arrow. Also remember that an equation is not balanced unless there is the same net charge on each side.

Self-Test

12. Balance the following by the oxidation-number-change method.

(a) $PH_3 + N_2O \longrightarrow H_3PO_4 + N_2$

(b) $NaIO_3 + Na_2SO_3 \longrightarrow Na_2SO_4 + NaI$

(c) $PbO_2 + HCl \longrightarrow PbCl_2 + H_2O + Cl_2$

New Terms

oxidation-number-change method

5.8 BALANCING REDOX EQUATIONS: THE ION-ELECTRON METHOD

Objectives

To learn a method of developing balanced net ionic equations for redox reactions in aqueous solution.

Review

Balancing a redox equation by this method is very simple if you remember to be sure to follow these steps:

1. divide the reaction into two half-reactions.
2. balance atoms other than H and O.
3. balance H and O (generally, O first, then H) using H^+ and H_2O for acid solution or H_2O and OH^- for basic solution.

4. count up the net charge on both sides of each half-reaction.
5. for each half-reaction add electrons to the most positive (least negative) to make the net charge on both sides of the arrow the same.
6. multiply the half-reactions by appropriate factors to make electron gain equal electron loss.
7. add the half-reactions.
8. cancel anything that appears the same on both sides of the equation.

If you wish to use the alternative method for basic solution explained on Page 148, use only H^+ and H_2O in Step 3.

When you use the ion-electron method it is extremely important to write the appropriate charges on each formula. If you mean H^+ and write H, without giving the charge, you will almost certainly get the number of electrons wrong. This, of course, will mean that you will multiply the half-reactions by the wrong factors and hence obtain an improperly balanced equation.

Self-Test

13. Balance the following by the ion-electron method. (All reactions in acid solution)

(a) $HNO_2 + I^- \longrightarrow I_2 + NO$

(b) $ClO_3^- + H_2S \longrightarrow Cl^- + S$

(c) $S_2O_8^{2-} + P \longrightarrow SO_4^{2-} + H_3PO_4$

14. Balance the following by the ion-electron method. (All reactions in basic solution)

(a) $CrO_4^{2-} + SO_3^{2-} \longrightarrow CrO_2^- + SO_4^{2-}$

(b) $HO_2^- + ClO_2 \longrightarrow ClO_2^- + O_2$

(c) $MnO_4^- + NO_2^- \longrightarrow MnO_2 + NO_3^-$

(d) $ClO^- + NH_3 \longrightarrow N_2H_4 + Cl^-$ (This reaction between bleach, OCl^-, and ammonia, NH_3, can produce poisonous hydrazine, N_2H_4. Be careful - don't mix household cleansers!)

New Terms

ion-electron method
half-reaction

5.9 QUANTITATIVE ASPECTS OF REACTIONS IN SOLUTION

Objectives

To express the concentration of solute in solution. You should learn how to solve problems dealing with the quantities of solutions used in chemical reactions.

Review

There are several ways of expressing concentration. Parts per hundred (percent) and parts per million (ppm) were discussed in the text. One of the most important concentration units is molarity, the ratio of moles of solute to liters of solution. If you know the number of moles of solute and the volume of the solution (in liters), the ratio gives molarity.

Example 5.2

What is the molarity of a solution containing 0.843 mole of NaCl in 750 ml of solution?

Solution

$$\text{molarity} = \frac{\text{moles}}{\text{liters}} = \frac{0.843 \text{ mole NaCl}}{0.750 \text{ liter solution}}$$

$$\text{molarity} = 1.12 \text{ M}$$

Molarity is a convenient conversion factor relating moles of solute to volume of solution. The molarity of the solution in Example 5.2 can be used to construct two conversion factors,

$$\frac{1.12 \text{ mole NaCl}}{1.00 \text{ liter}} \quad \text{and} \quad \frac{1.00 \text{ liter}}{1.12 \text{ mole NaCl}}$$

which can be used in calculations.

Example 5.3

What volume of 1.12 M NaCl contains 0.420 mole of NaCl?

Solution

$$0.420 \cancel{\text{mole NaCl}} \left(\frac{1.00 \text{ liter}}{1.12 \cancel{\text{mole NaCl}}}\right) = 0.375 \text{ liter}$$

The answer is 375 ml of solution.

Example 5.4

How many moles of NaCl are in 250 ml of 1.12 M NaCl solution?

Solution

$$250 \text{ ml} = 0.250 \text{ liter}$$

$$0.250 \text{ ~~liter~~} \left(\frac{1.12 \text{ mole NaCl}}{1.00 \text{ ~~liter~~}}\right) = 0.280 \text{ mole NaCl}$$

If you thoroughly understand the three problems above, as well as the example problems in the text, you are ready for the next Self-Test.

Self-Test

15. Calculate the molarity of the following solutions:

(a) 1.14 mole KI in 1.50 liter of solution ________

(b) 0.240 mole $CaCl_2$ in 500 ml of solution ________

(c) 3.50 g of NaCl in 0.0500 liter of solution ________

(d) 4.25 g $MgSO_4$ in 75.0 ml of solution ________

16. How many moles of $KClO_3$ are in 500 ml of 0.150 M solution? ________

17. How many moles of urea are in 250 ml of urine if the urea concentration is 0.32 M? ________

18. A normal adult excretes about 1500 ml of urine per day. If the urea concentration is 0.320 M, how many grams of urea are excreted per day? Urea has the formula, $CO(NH_2)_2$. ________

19. What volume of 0.250 M $CaCl_2$ are required to react completely with 300 ml of 0.150 M $AgNO_3$ according to the equation,

$$CaCl_2 + 2AgNO_3 \longrightarrow Ca(NO_3)_2 + 2AgCl$$ ________

20. How many moles of solid AgCl will be formed if 300 ml of 0.240 M $AgNO_3$ are added to 200 ml of 0.480 M HCl? The reaction is

$$AgNO_3 + HCl \longrightarrow AgCl + HNO_3$$ ________

New Terms

parts per million
molarity

5.10 EQUIVALENT WEIGHTS AND NORMALITY

Objectives

To define a quantity useful for dealing with the stoichiometry of acid-base and redox reactions and to extend this concept to reactions in solution.

Review

The equivalent weight (that is, the weight of one equivalent) is defined differently for redox reactions and for acid-base neutralization reactions.

Redox Reactions. The equivalent weight of a substance is simply its mole weight divided by the number of electrons that it gains or loses in the reaction. Thus when $KMnO_4$ is reduced to MnO_2 the manganese undergoes a change of 3 electrons.

$$\underset{+7}{KMnO_4} \longrightarrow \underset{+4}{MnO_2}$$

$$3e^-/KMnO_4$$

In this reaction the equivalent weight is the formula weight of $KMnO_4$ divided by 3.

$$\text{equiv wt } KMnO_4 = \frac{\text{mole weight}}{3} = \frac{158 \text{ g}}{3} = 52.7 \text{ g}$$

$$1 \text{ equiv } KMnO_4 = 52.7 \text{ g } KMnO_4$$

Note that we can also say that there are 3 equivalents of $KMnO_4$ per mole in this reaction.

How about a reaction in which $K_2Cr_2O_7$ is converted to $CrCl_3$? What is the equivalent weight of $K_2Cr_2O_7$? In the reaction each Cr gains 3 electrons; since $K_2Cr_2O_7$ has two Cr atoms, each $K_2Cr_2O_7$ gains 6 electrons.

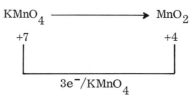

$$K_2Cr_2O_7 \longrightarrow 2CrCl_3$$

$$+6 \qquad\qquad\qquad +3$$

$$6e^-/K_2Cr_2O_7$$

The equivalent weight of $K_2Cr_2O_7$ is its formula weight divided by 6. There are six equivalents of $K_2Cr_2O_7$ per mole.

Acid-Base Reactions. For an acid the equivalent weight is equal to the formula weight divided by the number of H^+ ions neutralized. If H_3PO_4 is completely neutralized to give PO_4^{3-}, the equivalent weight of H_3PO_4 is the formula weight divided by 3. On the other hand, if H_3PO_4 is only partially neutralized to give HPO_3^{2-}, only two of the three H^+ ions are neutralized and the equivalent weight is the formula weight divided by 2.

Remember that the whole rationale for equivalent weights is that one equivalent of one substance reacts exactly with one equivalent of another.

Redox

1 equivalent of oxidizing agent reacts with 1 equivalent of reducing agent.

Acid-Base

1 equivalent of acid reacts exactly with 1 equivalent of base.

Normality. Remember that normality is similar to molarity, except that the units are equivalents/liter. Also remember the useful relationship between normality (N) and molarity (M),

$$N = n \cdot M$$

where n is the number by which the formula weight must be divided to give equivalent weight (the number of electrons transferred in redox, or the number of H^+ or OH^- neutralized in acid-base reactions).

When solutions are reacted, remember that

$$N_1V_1 = N_2V_2$$

where N_1 and V_1 are the normality and volume of solution 1; N_2 and V_2 are the normality and volume of solution 2.

Example 5.5

How many ml of 0.180 N $Ba(OH)_2$ are required to react completely with 22.0 ml of 0.0800 N HCl?

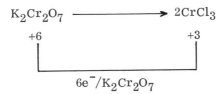

Solution

	acid	base
N	0.0800	0.180
V	22.0 ml	x

$$N_1V_1 = N_2V_2$$

$$(0.0800\ N)(22.0\ ml) = (0.180\ N)(x)$$

$$x = \frac{(0.0800\ N)(22.0\ ml)}{(0.180\ N)} = 9.78\ ml$$

The volume of $Ba(OH)_2$ solution required is 9.78 ml.

Self-Test

21. Calculate the equivalent weight for the following substances. The products produced when they react are given in parenthesis.

(a) $NaBiO_3$ (Bi^{3+}) ________

(b) $NaIO_3$ (I^-) ________

(c) H_2SO_4 (SO_4^{2-}) ________

(d) H_2SO_4 (HSO_4^-) ________

(e) $Na_2S_2O_3$ ($S_4O_6^{2-}$) ________

22. $Na_2S_2O_4$ reacts with $CuSO_4$ in solutions containing ammonia to produce SO_3^{2-} and metallic copper. How many grams of $Na_2S_2O_4$ are needed to completely reduce 15.0 g of $CuSO_4$?

____

23. How many grams of K_2CrO_4 are needed to prepare 300 ml of 0.100 N solution if the solution will be used in a reaction in which the Cr is reduced to the +3 oxidation state?

____

24. How many ml of 0.100 N HCl react with 19.5 ml of 0.220 N KOH?

____

25. What is the normality of an H_2SO_4 solution if 27.0 ml of the solution is required to neutralize 14.3 ml of 0.35 N NaOH?

____

New Terms

equivalent
equivalent weight
normality

5.11 CHEMICAL ANALYSIS

Objectives

To learn how the principles discussed in the previous sections of this chapter can be put to practical use for the purposes of chemical analysis.

Review

There are three important ideas developed in this section. The first of these involves analyzing a substance by forming and weighing a compound of known composition containing all of one component of the original sample. For example, if a mixture is known to contain silver, the compound AgCl can be formed by adding Cl^- to a solution of a weighed sample of the mixture. The insoluble AgCl is filtered, dried and weighed. From the known composition of AgCl, the weight of Ag in the precipitate can be calculated. Since the Ag came from the original sample, the weight of Ag in the sample is known. The practical application of this method, of course, depends on a knowledge of solubilities.

Example 5.6

A 1.00-g sample of an ore known to contain silver was dissolved in HNO_3 and treated with Cl^-. 0.275 g of AgCl was obtained. What was the percent Ag in the ore?

Solution

From the weight of AgCl we calculate the weight of Ag.

$$0.275 \text{ g AgCl} \left(\frac{107.9 \text{ g Ag}}{143.4 \text{ g AgCl}} \right) \quad 0.207 \text{ g Ag}$$

(If you don't understand this calculation, review Section 2.3) All of this 0.207 g of Ag came from the ore. Therefore, the percent silver in the ore is

$$\% \text{ Ag} = \frac{0.207 \text{ g}}{1.00 \text{ g}} \times 100 = 20.7\%$$

The second major point in this section involves the use of the quantitative relationships of solution stoichiometry in a procedure called titration. One reactant in solution is added to another from a calibrated buret. If the concentration of the titrant (the solution delivered from the buret) is known, the amount of the reactant in the other solution can be calculated. Review Examples 5.18 and 5.19 in the text.

The third item discussed in this section is dilution. Remember the two simple relationships,

$$N_i V_i = N_f V_f$$

$$M_i V_i = M_f V_f$$

Self-Test

26. A 0.833-g sample of a mixture of NaCl and $CaCl_2$ was dissolved in water and treated with Na_2CO_3 to precipitate $CaCO_3$. The precipitate was filtered and dried and found to weigh 0.415 g. What percent of the original sample was $CaCl_2$?

    ____

27. A 0.400-g sample of an alloy of iron and nickel was dissolved in HCl to give Fe^{2+} and Ni^{2+}. The resulting solution was titrated with 22.3 ml of 0.200 N $KMnO_4$, causing Fe^{2+} to be oxidized to Fe^{3+}. What percent of the alloy is iron?

    ____

28. How many ml of 3.00 M HCl must be used to prepare 500 ml of 0.100 M HCl?

    ____

29. How much water must be added to 50.0 ml of 1.00 N NaOH to produce 0.100 N NaOH?

    ____

New Terms

volumetric analysis
titration
titrant
indicator
end point
equivalence point

ANSWERS TO SELF-TEST QUESTIONS

1. a, d
2. (a) $NH_3 + H_2O \rightleftharpoons NH_4^+ + OH^-$

(b) $HCN + H_2O \rightleftharpoons H_3O^+ + CN^-$

3. (a) S, rules 1 and 3 (b) S, rule 5 (c) I, rule 4 (d) I, rule 8 (e) I, rule 6 (f) I, rule 5 (g) I, rule 7 (h) S, rule 2
4. (a) $Na^+ + I^- + Ag^+ + NO_3^- \longrightarrow AgI + Na^+ + NO_3^-$

   $Ag^+ + I^- \longrightarrow AgI$

   (b) $Pb^{2+} + 2NO_3^- + Ba^{2+} + 2OH^- \longrightarrow Pb(OH)_2 + Ba^{2+} + 2NO_3^-$

   $Pb^{2+} + 2OH^- \longrightarrow Pb(OH)_2$

   (c) $AgCl + Na^+ + Br^- \longrightarrow AgBr + Na^+ + Cl^-$

   $AgCl + Br^- \longrightarrow AgBr + Cl^-$

   (d) $Zn^{2+} + 2Cl^- + 2Na^+ + CO_3^{2-} \longrightarrow 2Na^+ + 2Cl^- + ZnCO_3$

   $Zn^{2+} + CO_3^{2-} \longrightarrow ZnCO_3$

   (e) $Ba^{2+} + 2Cl^- + 2Na^+ + 2Br^- \longrightarrow Ba^{2+} + 2Br^- + 2Na^+ + 2Cl^-$

   (no net reaction)
5. (a) $CaCO_3 + 2H^+ + 2Cl^- \longrightarrow H_2O + CO_2 + Ca^{2+} + 2Cl^-$

   $CaCO_3 + 2H^+ \longrightarrow H_2O + CO_2 + Ca^{2+}$

   (b) $2NH_4^+ + SO_4^{2-} + 2Na^+ + 2\,OH^- \longrightarrow 2NH_3 + 2H_2O + 2Na^+ + SO_4^{2-}$

   $2NH_4^+ + 2\,OH^- \longrightarrow 2NH_3 + 2H_2O$; dividing through by 2 gives

   $NH_4^+ + OH^- \longrightarrow NH_3 + H_2O$

   (c) $2Na^+ + C_2O_4^{2-} + 2H^+ + 2Cl^- \longrightarrow 2Na^+ + 2Cl^- + H_2C_2O_4$

   $C_2O_4^{2-} + 2H^+ \longrightarrow H_2C_2O_4$

   (d) $BaCO_3 + 2H^+ + SO_4^{2-} \longrightarrow BaSO_4 + H_2O + CO_2$

   (net is same as above)

   (e) $2K^+ + SO_3^{2-} + 2H^+ + SO_4^{2-} \longrightarrow 2K^+ + SO_4^{2-} + H_2O + SO_2$

   $SO_3^{2-} + 2H^+ \longrightarrow SO_2 + H_2O$
6. (a) $NiCl_2 + Na_2CO_3 \longrightarrow 2NaCl + NiCO_3$

   $Ni^{2+} + 2Cl^- + 2Na^+ + CO_3^{2-} \longrightarrow 2Na^+ + 2Cl^- + NiCO_3(s)$

   $Ni^{2+} + CO_3^{2-} \longrightarrow NiCO_3(s)$

   (b) $2NaCl + CaBr_2 \longrightarrow 2NaBr + CaCl_2$

   $2Na^+ + 2Cl^- + Ca^{2+} + 2Br^- \longrightarrow 2Na^+ + 2Br^- + Ca^{2+} + 2Cl^-$

   (no net reaction)

   (c) $MgCl_2 + 2NaOH \longrightarrow Mg(OH)_2 + 2NaCl$

   $Mg^{2+} + 2Cl^- + 2Na^+ + 2OH^- \longrightarrow Mg(OH)_2(s) + 2Na^+ + 2Cl^-$

$Mg^{2+} + 2\,OH^- \longrightarrow Mg(OH)_2(s)$

(d) $Na_2SO_4 + Mg(NO_3)_2 \longrightarrow 2NaNO_3 + MgSO_4$

$2Na^+ + SO_4^{2-} + Mg^{2+} + 2NO_3^- \longrightarrow 2Na^+ + 2NO_3^- + Mg^{2+} + SO_4^{2-}$

(no net reaction)

(e) $Na_2S + 2HCl \longrightarrow 2NaCl + H_2S$

$2Na^+ + S^{2-} + 2H^+ + 2Cl^- \longrightarrow 2Na^+ + 2Cl^- + H_2S(g)$

$S^{2-} + 2H^+ \longrightarrow H_2S(g)$

7. (a) $Ca(OH)_2 + 2HCl \longrightarrow CaCl_2 + 2H_2O$

(b) $2NH_3 + H_2SO_4 \longrightarrow (NH_4)_2SO_4$

(c) $2Al(OH)_3 + 3H_2SO_4 \longrightarrow Al_2(SO_4)_3 + 6H_2O$

(d) $NaOH + HNO_3 \longrightarrow NaNO_3 + H_2O$

(e) $MgO + H_2SO_4 \longrightarrow MgSO_4 + H_2O$

8. (a) CaC_2O_4, $Ca(HC_2O_4)_2$ (b) $CaCO_3$, $Ca(HCO_3)_2$

(c) $Ca_3(AsO_4)_2$, $CaHAsO_4$, $Ca(H_2AsO_4)_2$ (d) $CaBr_2$

9. One possible set of reactants for each is given here. However, there is more than one way to "skin a cat". If you have chosen different reactants, ask your teacher if they are satisfactory.

(a) $FeCl_3(aq) + 3AgNO_3(aq) \longrightarrow Fe(NO_3)_3(aq) + 3AgCl(s)$

(b) $AgNO_3(aq) + KBr(aq) \longrightarrow AgBr(s) + KNO_3(aq)$

in general: $Ag^+ + Br^- \longrightarrow AgBr(s)$

(c) $Ba(OH)_2(aq) + MgBr_2(aq) \longrightarrow BaBr_2(aq) + Mg(OH)_2(s)$

(d) $Ba(OH)_2(aq) + Na_2SO_4(aq) \longrightarrow BaSO_4(s) + 2NaOH(aq)$

10. (a) $CuCl_2(aq) + 2NaOH(aq) \longrightarrow Cu(OH)_2(s) + 2NaCl(aq)$

$Cu(OH)_2(s) + 2H_2SO_4 \longrightarrow Cu(HSO_4)_2(aq) + 2H_2O$

(b) $NiCl_2(aq) + 2NaOH(aq) \longrightarrow 2NaCl(aq) + Ni(OH)_2(s)$

$Ni(OH)_2(s) + H_2SO_4(aq) \longrightarrow NiSO_4(aq) + 2H_2O$

(c) $MgCl_2(aq) + 2NaOH(aq) \longrightarrow Mg(OH)_2(s) + 2NaCl$

$Mg(OH)_2(s) + 2HNO_3(aq) \longrightarrow Mg(NO_3)_2(aq) + 2H_2O$

(d) $CaO(s) + 2HNO_3(aq) \longrightarrow Ca(NO_3)_2(aq) + H_2O$

11. (a) $FeCO_3(aq) + 2HCl(aq) \longrightarrow FeCl_2(aq) + H_2O + CO_2(aq)$

(b) $CoCO_3(aq) + 2HNO_3(aq) \longrightarrow Co(NO_3)_2(aq) + H_2O + CO_2(g)$

(c) $NH_4NO_2(aq) + NaOH(aq) \longrightarrow NaNO_2(aq) + H_2O + NH_3(g)$

(d) $ZnS(s) + 2HCl(aq) \longrightarrow ZnCl_2(aq) + H_2S(g)$

(e) $CaCl_2(aq) + 2NaOH(aq) \longrightarrow CaCO_3(s) + 2NaCl(aq)$

$CaCO_3(s) + 2HNO_3(aq) \longrightarrow Ca(NO_3)_2(aq) + H_2O + CO_2(g)$

12. (a) $PH_3 + 4N_2O \longrightarrow H_3PO_4 + 4N_2$

(b) $NaIO_3 + 3Na_2SO_3 \longrightarrow 3Na_2SO_4 + NaI$

(c) $PbO_2 + 4HCl \longrightarrow PbCl_2 + 2H_2O + Cl_2$

13. (a) $2H^+ + 2HNO_2 + 2I^- \longrightarrow 2NO + I_2 + 2H_2O$

(b) $ClO_3^- + 3H_2S \longrightarrow Cl^- + 3S + 3H_2O$

(c) $8H_2O + 5S_2O_8^{2-} + 2P \longrightarrow 10\,SO_4^{2-} + 2H_3PO_4 + 10\,H^+$

14. (a) $H_2O + 2CrO_4^{2-} + 3SO_3^{2-} \longrightarrow 2CrO_2^- + 3SO_4^{2-} + 2\,OH^-$

(b) $OH^- + HO_2^- + 2ClO_2 \longrightarrow 2ClO_2^- + O_2 + H_2O$

(c) $H_2O + 2MnO_4^- + 3NO_2^- \longrightarrow 2MnO_2 + 3NO_3^- + 2\,OH^-$

(d) $ClO^- + 2NH_3 \longrightarrow N_2H_4 + Cl^- + H_2O$

15. (a) 0.760 M (b) 0.480 M (c) 1.20 M (d) 0.471 M

16. 0.0759 mole $KClO_3$ 17. 0.080 mole urea 18. 28.8 g of urea

19. 90.0 ml 20. 0.0720 mole AgCl (HCl is in excess)

21. (a) 140 g $NaBiO_3$ (b) 33.0 g $NaIO_3$ (c) 49.0 g H_2SO_4
(d) 98.1 g H_2SO_4 (e) 158 g $Na_2S_2O_3$

22. 16.4 g $Na_2S_2O_4$ (eq.wt. $Na_2S_2O_4$ = 87.06 g; $CuSO_4$ = 79.8 g)

23. 1.94 g K_2CrO_4 24. 42.9 ml 25. 0.185 N 26. 55.3% $CaCl_2$

27. 62.3% Fe 28. 16.7 ml of 3.00 M HCl

29. 450 ml water added (total volume = 500 ml)

6 GASES

In this chapter we examine the physical and chemical behavior of gases. This includes the way the properties of a gas depend on pressure, volume, and temperature. We will see that it is a relatively simple matter to measure the molecular weights of gaseous substances. The study of gases has also led to knowledge about the microscopic behavior of gases, including information about molecular size and the kinds of attractive forces that exist between gaseous atoms and molecules.

6.1 VOLUME AND PRESSURE

Objectives

To learn how pressure and volume are defined. You should learn how the pressure of a gas is measured and the units in which pressure is expressed.

Review

Gases expand to fill whatever container they are placed in. The volume of a gas is therefore the volume of its container.

Pressure is force per unit area. In the English system pressure can be expressed in pounds per square inch (psi). The pressure of a gas is usually given in terms of the height of a mercury column that exerts the same pressure as the gas. Remember that a mercury column 1 mm high exerts a pressure of 1 torr. At sea level, the pressure exerted by the atmosphere fluctuates about 760 torr. One standard atmosphere (1 atm) is defined as precisely 760 torr. The SI unit of pressure is the pascal (Pa).

A standard atmosphere = 1.013×10^5 Pa. You should be able to convert pressures from torr to atm, and vice versa using the relationship, 1 atm = 760 torr.

Gas pressures are measured using manometers. For an open-end manometer, you see in the text that the equation used to calculate the gas pressure depends on whether the gas pressure is greater than or less than the atmospheric pressure. Don't try to memorize these equations - it's too easy to get them confused. It is much better if you can learn how to analyze the manometer. Remember that the lower of the two liquid levels is always chosen as the reference level. At this level the pressures on both the left and right sides are the same. Figure out what they are, equate them, and then solve for the pressure of the gas.

Sometimes a liquid other than mercury is used in a manometer. To express the pressure in torr it is necessary to convert the difference in heights in the liquid columns to the difference in heights that would have been found had mercury been in the manometer. This is obtained by multiplying the difference in heights of the liquid by a ratio of the densities of the liquid and mercury. Mercury is the most dense liquid ever used in a manometer. Therefore, the height of the mercury column will always be less than the height of the liquid column. Always set up the ratio of densities so that the calculated equivalent column of mercury is less than that of the liquid. Review Example 6.1 in the text and Example 6.1 below.

Example 6.1

A liquid having a density of 1.22 g/ml is used in a manometer. In measuring the pressure of a gas a difference of 15.6 cm was observed between the liquid levels in the two arms. What is the mercury equivalent, in torr, of this column of liquid. The density of mercury is 13.6 g/ml.

Solution

The mercury equivalent of the liquid column is obtained by multiplying the height of the liquid column (15.6 cm) by a ratio of densities.

$$P_{Hg} = P_{liquid} \times \text{(ratio of densities)}$$

Since P_{Hg} will be less than P_{liquid}, the ratio of densities must have some value smaller than 1, so that when P_{liquid} is multiplied by this ratio a smaller value is obtained. A value less than 1 is only obtained as 1.22/13.6. Therefore,

$$P_{Hg} = 15.6 \text{ cm} \left(\frac{1.22}{13.6}\right) = 1.40 \text{ cm Hg}$$

Since 1 torr = 1 mm Hg,

$$P_{Hg} = 14.0 \text{ mm Hg} = 14.0 \text{ torr}$$

Self-Test

1. If a gas exerts a pressure of 1.35 atm, what is its pressure in torr?

   ____

2. If a gas exerts a pressure of 630 torr, what is its pressure in atm?

   ____

3. A gas in a container is attached to an open-end manometer filled with mercury. The mercury level in the arm connected to the container is 18.5 mm below the level in the open arm. The atmospheric pressure is 755.3 torr. Calculate the pressure of the gas. (It will help if you sketch a picture of the apparatus.)

   ____

4. A gas contained in a vessel exerts a pressure of 1.14 atm. The vessel is connected to an open-end manometer filled with mercury. The atmospheric pressure is 766 torr. What will be the difference in heights (measured in cm) between the levels of mercury in the two arms?

   ____

5. Two gases in the cylinders shown in Figure 6.1 exert pressures of 785 torr and 790 torr. If the manometer is filled with water (d = 1.00 g/ml), what will be the difference in heights between the liquid levels in the two arms?

   ____

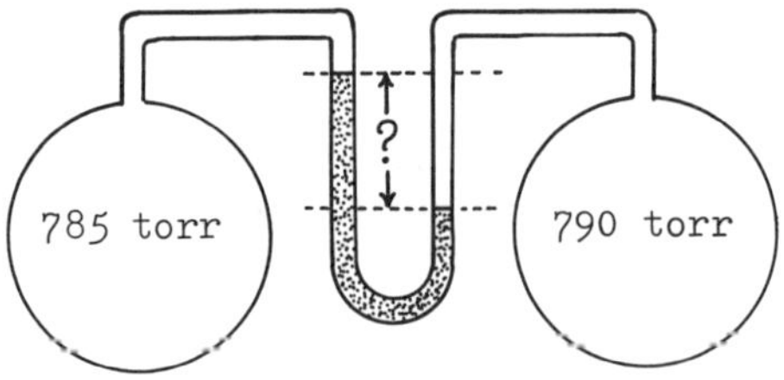

Figure 6.1

6. Calculate the pressure, in atm, exerted on a skin diver at a depth of 12 fathoms on a day when the atmospheric pressure is 752 torr (a barometric pressure of 29.6 inches of mercury). The density of sea water is 1.02 g/ml; 1 fathom equals 6 feet.

   ____

New Terms

barometer
standard atmosphere
pascal
torr
manometer

6.2 BOYLE'S LAW

Objectives

To learn how the volume of a gas varies with the pressure exerted on it.

Review

Remember that as the pressure on a gas increases, the gas is squeezed into a smaller volume. Boyle's law says that at constant temperature the product, PV, for a fixed quantity of gas is constant. For real gases this is not quite true, although near atmospheric pressure and room temperature most gases follow Boyle's law quite well. An ideal gas is a hypothetical gas that would obey Boyle's law perfectly under all conditions.

You will be expected to be able to perform calculations dealing with Boyle's law. There are two ways to approach these problems:

(1) You can memorize the equation, $P_1V_1 = P_2V_2$. This equation can be used, for example, to calculate the final pressure of a gas if you know its initial pressure and volume, and its final volume.

(2) You can use the idea that the final volume of a gas is equal to its initial volume multiplied by a ratio of pressures. For example, an increase in pressure causes the final volume to be smaller than the initial volume. The ratio of pressures must therefore be smaller than one so that when the initial volume is multiplied by the ratio the result is a smaller volume. Since this reasoning approach requires some practice, another example is presented below.

Example 6.2

A gas, initially occupying 2.00 liters at 780 torr, is placed in a container (at the same temperature) in which its pressure is 740 torr. What is the volume of the container?

Solution

Remember to set up a table of initial and final conditions to avoid confusion.

	initial	final
P	780 torr	740 torr
V	2.00 liters	?

Next, reason through the problem. The pressure on the gas has decreased (780 → 740). Therefore, the gas must have expanded. This means that the final volume must be larger than 2.00 liters. Which ratio of pressures do we choose to multiply the initial volume by?

$$\frac{780 \text{ torr}}{740 \text{ torr}} \quad \text{or} \quad \frac{740 \text{ torr}}{780 \text{ torr}}$$

Obviously (780/740) is the correct choice since it is the only one that will make the final volume larger. The answer is then obtained as

$$V_f = 2.00 \text{ liters} \left(\frac{780 \text{ torr}}{740 \text{ torr}}\right) = 2.11 \text{ liters}$$

As a final note, remember that is is absolutely necessary to have the units of the two quantities (P or V) in a ratio the same. The units in the ratio must cancel.

Self-Test

7. A gas at 745 torr occupies 250 ml. What volume will it occupy at 300 torr if the temperature remains the same?

   ____

8. A gas occupies a volume of 180 ml at a pressure of 450 torr. If the gas is transferred to a 2.00-liter container at the same temperature, what will be the new pressure?

   ____

9. Gasoline vapor mixed with air at atmospheric pressure (1 atm) is drawn into a 610-ml cylinder in an auto engine. Before the mixture is exploded the piston compresses the gas to a volume of 73.1 ml. What is the pressure of the gas when the mixture is ignited?

   ____

New Terms

Boyle's law
ideal gas

6.3 CHARLES' LAW

Objectives

To learn how the volume of a gas is related to its temperature.

Review

Remember that absolute zero occurs at $-273^{o}C$. Temperatures used in gas law calculations are always expressed in kelvin. The kelvin temperature is obtained by adding 273 to the Celsius temperature.

$$K = {}^{o}C + 273$$

Reference conditions for gases have been chosen to be $0^{o}C$ (standard temperature) and 760 torr (standard pressure). STP is the abbreviation used to specify standard temperature and pressure.

Calculations involving Charles' law can also be approached in two ways:

(1) You can use the equation,

$$\frac{V_1}{T_1} = \frac{V_2}{T_2}$$

(2) You can use the fact that as the temperature of a gas increases, at constant pressure, its volume increases. Stated more simply, the gas expands as it gets hot.

As before we have a situation where the final volume is equal to the initial volume multiplied by a ratio of absolute temperatures.

Combined gas law calculations, involving Boyle's law and Charles' law together can be solved using the reasoning approach; for example,

$$V_f = V_i\text{(ratio of pressures)(ratio of temepratures)}$$

You can also use the equation,

$$\frac{P_1V_1}{T_1} = \frac{P_2V_2}{T_2}$$

Review Examples 6.3 and 6.4 in the text.

Self-Test

10. Convert the following temperatures to kelvin.

(a) $14^{\circ}C$ ________ (b) $35^{\circ}C$ ________ (c) $-30^{\circ}C$ ________

11. Convert the following temperatures to degrees Celsius.

(a) 315 K ________ (b) 298 K ________ (c) 77 K ________

12. A gas at $33^{\circ}C$ occupies 14.0 liters. What volume would it occupy at $66^{\circ}C$ if the pressure remains the same? ________

13. To what Celsius temperature must a gas, initially at $30^{\circ}C$, be heated in order to triple its volume, if the pressure remains constant? ________

14. A gas at $25^{\circ}C$ and 740 torr occupies 840 ml. What volume will it occupy at $14^{\circ}C$ and 650 torr? ________

15. A sealab having an open hatch at the bottom is submerged to a depth at which the pressure is 8.5 atm and the temperature is $13^{\circ}C$. At the surface (P = 1 atm) its volume is 5000 ft^3 and the temperature is a balmy $30^{\circ}C$. If no air is pumped into the sealab, how much usable air space will remain when it is submerged? ________

New Terms

absolute zero
kelvin temperature scale
absolute temperature scale
Charles' law
standard temperature
standard pressure
STP

6.4 DALTON'S LAW OF PARTIAL PRESSURES

Objectives

To observe how gases behave in mixtures.

Review

Dalton's law is very simple. Each gas in a mixture behaves independently of any other gases present and exerts a pressure (called its partial pressure) that is the same as it would exert if it were alone in the container. The total pressure of the mixture is simply the sum of the pressures of each of the gases.

A practical application of Dalton's law is the calculation of the pressure of a gas collected over a liquid such as water. Review Examples 6.5 and 6.6 in the text before attempting the Self-Test below.

Self-Test

16. 300 ml of argon at a pressure of 420 torr and 300 ml of helium at 240 torr are placed into the same 300-ml container. The temperatures of the separate gases and the mixture are the same. What is the partial pressure of each gas in the mixture and the total pressure of the mixture?

____

17. 200 ml of N_2 at 30°C and 750 torr is mixed with some O_2 and transferred to a 500-ml container at 30°C. The total pressure of the mixture is found to be 680 torr. What is the partial pressure of each gas in the mixture?

____

18. 400 ml of oxygen was collected over water at 25°C at a total pressure of 765 torr. What is the partial pressure of the trapped oxygen?

____

New Terms

partial pressure
Dalton's law of partial pressure
vapor pressure

6.5 LAWS OF GAY-LUSSAC

Objectives

To learn the relationship that exists between the pressure of a gas and its temperature. You should also learn that the volume of a gas is directly proportional to the number of moles of gas.

Review

Remember that as the temperature of a fixed volume of gas is raised the pressure of the gas increases. Problems dealing with pressure-temperature changes at constant volume can be solved either with the equation,

$$\frac{P_1}{T_1} = \frac{P_2}{T_2}$$

or by the reasoning approach that we have been following throughout this chapter. This is illustrated in Example 6.7 in the text.

Gay-Lussac's law of comgining volumes states that at constant temperature and pressure the volumes of gases consumed and/or produced in a chemical reaction are related to one another as ratios of small whole numbers. In fact, under conditions of constant T and P, the volume ratios are the same as the ratios of the coefficients in the balanced chemical equation.

Also of historical importance is Avogadro's principle - equal volumes of gas at the same T and P contain equal numbers of molecules. This can also be expressed as the volume of a gas at constant temperature and pressure is proportional to the number of moles of gas present.

Self-Test

19. A tire is inflated to a pressure of 28 psi at 25°C. The tire is driven for several hours and heats up to 40°C. What is the air pressure in the tire at this higher temperature? __________

20. What volume of O_2 is required to completely react with 150 ml of C_2H_6 (both volumes measured at the same T and P) according to the equation, $2C_2H_6(g) + 7\,O_2(g) \longrightarrow 4CO_2(g) + 6H_2O$? __________

21. What volume of air (composed of 20% O_2 by volume) at 25°C and 1.00 atm are required to react with 500 ml of C_2H_6 (measured at 30°C and 850 torr)? (This problem requires several steps - think about how to solve the problem before working with the numbers.) __________

New Terms

law of Gay-Lussac
Gay-Lussac's law of combining volumes
Avogadro's principle

6.6 THE IDEAL GAS LAW

Objectives

To obtain an equation that encompasses all of the gas laws that we have examined in previous sections. You should learn the volume occupied by one mole of a gas at STP and you should learn to solve problems dealing with molar quantities of gas.

Review

You should learn the ideal gas law,

$$PV = nRT$$

Remember that at STP one mole of a gas occupies a volume of approximately 22.4 liters (the molar volume at STP).

The gas constant, R, can have different numerical values depending on the units used to express pressure and volume. The value used most frequently in the text is 0.0821 liter atm/mole K. If you learn this value, it's important that you also learn the units that go with it. If you use this value in the ideal gas law, remember that the pressure must be expressed in atm and the volume in liters; otherwise incorrect numerical answers will result regardless of how good you are at arithmetic.

A very important application of the ideal gas law is in the determination of molecular weights of gaseous substances. Review Examples 6.12 and 6.13 in the text as well as the Example below.

Example 6.3

A sample of a gas was found to have a density of 1.64 g/liter at 30°C and 0.930 atm. What is the molecular weight of the gas?

Solution

When you are asked to compute a molecular weight, you actually are being asked to calculate the number of grams per mole. The necessary data for this are P, V, T and a weight of gas (in grams). From the P, V, T data you can calculate the number of moles of gas from the ideal gas law. The molecular weight is then obtained simply by taking the ratio of grams of gas/moles of gas.

The density in this question gives the weight of one liter of gas. Therefore, we have: P = 0.930 atm, V = 1.00 liter, T = 273 + 30 = 303 K, mass = 1.64 g. Solving the ideal gas law for n,

$$n = \frac{PV}{RT} = \frac{(0.930 \text{ atm})(1.00 \text{ liter})}{(0.0821 \text{ liter atm/mole K})(303 \text{ K})}$$

$$n = 0.0374 \text{ mole}$$

Then,

$$\text{M.W.} = \frac{1.64 \text{ g}}{0.0374 \text{ mole}} = 43.9 \text{ g/mole}$$

Self-Test

22. Calculate the numerical value of R having the units,

    (a) ml atm/mole K ______________________

    (b) liter torr/mole K ______________________

23. How many moles of gas are present in 3.00 liters at 800 torr and 40^oC?

    ____

24. Calculate the pressure in atm exerted by 0.10 mole of argon at -20^oC in a 10-liter container.

    ____

25. 0.625 g of an unknown gas occupies 500 ml at STP. What is its molecular weight?

    ____

26. What volume would 28 g of O_2 occupy at 800 torr and 27^oC?

    ____

27. Calculate the density of N_2 at STP. ______________________

28. Butane, from a cigarette lighter, has a density of 2.30 g/liter at 22^oC and 730 torr. What is the molecular weight of butane?

    ____

    If its empirical formula is C_2H_5, what is its molecular formula?

    ____

New Terms

ideal gas law

perfect gas law

equation of state for an ideal gas

molar volume

6.7 GRAHAM'S LAW OF EFFUSION

Objectives

To learn how the rates of effusion (and diffusion) of gases are related to their molecular weights.

Review

Diffusion and effusion are similar, although not identical, processes. Both refer to the rates at which gas molecules move from one place to another. Graham's law is summarized in the equation,

$$\frac{\text{rate of effusion of gas (A)}}{\text{rate of effusion of gas (B)}} = \sqrt{\frac{M_B}{M_A}}$$

where M_A and M_B are the molecular weights of A and B, respectively.

Self-Test

29. Which of the following molecules diffuses faster?

    (a) H_2O or H_2S __________

    (b) NH_3 or H_2O __________

    (c) CO_2 or NO_2 __________

30. Calculate the ratio of the rates of effusion, R_{NO}/R_{NO_2}.

    ____

31. How many times faster does $^{235}UF_6$ diffuse than $^{238}UF_6$?

    ____

32. What would the molecular weight of a gas be if it diffuses only one sixth as fast as H_2? __________

New Terms

diffusion
effusion
Graham's law

6.8 THE KINETIC MOLECULAR THEORY

Objectives

To learn and understand the postulates of the theory that was developed to explain the gas laws which were discussed in earlier sections.

Review

The kinetic molecular theory was developed to explain why gases behave the way they do. The theory is described by a set of postulates presented in detail in the text. In summary, the kinetic molecular theory (or simply the kinetic theory) views a gas as being composed of very tiny molecules, having negligible volume themselves, separated by very large distances from one another. The molecules are in rapid random motion, colliding with the walls of the container and with each other. The pressure exerted by a gas results from collisions of the molecules with the walls. It is further postulated that the molecules do not attract each other and that there is a distribution of molecular speeds, ranging from very slow molecules to extremely fast ones. Associated with the distribution of molecular speeds there is a corresponding distribution of kinetic energies. To account for Graham's law it is necessary to postulate that the average kinetic energy of a gas depends only on the absolute temperature. This is perhaps the most important aspect of the kinetic theory because it applies to any collection of molecules. At a given temperature the average kinetic energy is the same for any collection of molecules, regardless of their chemical makeup or whether they are in a gas, a liquid, or a solid.

Self-Test

33. (Multiple choice) Kinetic theory accounts for Boyle's law by postulating that

(a) the average kinetic energy depends only on temperature
(b) a gas is mostly empty space
(c) there are no attractive forces between molecules
(d) the molecules are in rapid random motion ________

34. (Multiple choice) To account for Charles' law it is necessary to postulate that

(a) there is a distribution of molecular speeds between gas molecules
(b) the molecules have negligibly small volumes
(c) molecules collide with each other
(d) on the average, molecules move faster at higher temperature

____

New Terms

Brownian motion
mole fraction

6.9 DISTRIBUTION OF MOLECULAR SPEEDS

Objectives

To learn more about the way the distribution of molecular speeds and molecular kinetic energies depends on the absolute temperature.

Review

The focal point of this section is Figure 6.12 in the text. There are several features of this figure you should learn well.

(1) At any temperature the fraction of molecules having zero kinetic energy is zero.

(2) The fraction of molecules having very large kinetic energies gradually approaches zero at high kinetic energies.

(3) The maximum on the curve at a given temperature represents the kinetic energy possessed by the largest fraction of molecules. This is termed the most probable kinetic energy since it is the one most likely to be found if the molecules were sampled at random.

(4) At successively higher temperatures the height of the maximum decreases. The total area under the curve represents the sum of all of fractions, and must equal 1.00. Since the curve gets higher at large kinetic energies, it must get lower elsewhere so that the area remains constant.

(5) At any given temperature the average kinetic energy occurs at a slightly higher kinetic energy than the most probable kinetic energy. This is a consequence of the unsymmetrical shape of the distribution curve.

(6) The average kinetic energy increases as the temperature increases.

Self-Test

35. After studying this section, sketch the kinetic energy distribution for a gas at two different temperatures on the axes which follow. Indicate the average and most probable kinetic energies at both temperatures.

New Terms

most probable kinetic energy

6.10 REAL GASES

Objectives

To see how the postulates of the kinetic theory must be modified to account for the properties of real gases, which do not obey the ideal gas law exactly.

Review

There are two defects in the kinetic theory presented in Section 6.8.

(1) Molecules do have attractive forces between them. These are either dipole-dipole attractions between polar molecules or London forces between nonpolar molecules.

(2) Molecules themselves do have a finite volume which is not negligible compared to the total volume when the molecules are squeezed close together.

The van der Waals equation,

$$\left(P + \frac{n^2a}{V^2}\right)\left(V - nb\right) = nRT$$

attempts to apply corrections to the pressure and volume of a gas in the ideal gas law. The actual pressure and volume are modified to give a pressure and volume that the gas would have if it were an ideal gas (i.e., if there were no attractive forces and if the gas molecules had zero volume). In the equation, the constant a is proportional to the strengths of the attractive forces and b is proportional to the size of the molecules.

Self-Test

36. Use the data in Table 6.3 to answer the following:

(a) Which gas has the larger molecules, CH_4 or H_2O? ________

(b) Which gas has the greater attractive forces between molecules, NH_3 or H_2O? ________

(c) Which gas has the larger molecules, H_2O or C_2H_5OH? ________

(d) Which gas has the greater attractive forces, O_2 or CH_4? ________

New Terms

excluded volume
van der Waals equation of state

ANSWERS TO SELF-TEST QUESTIONS

1. 1026 torr 2. 0.829 atm 3. 773.8 torr 4. 100 mm 5. 68 mm 6. 3.16 atm (2.17 atm from sea water, 0.99 atm from atmosphere) 7. 621 ml 8. 40.5 torr 9. 8.34 atm 10. (a) 287 K (b) 308 K (c) 243 K 11. (a) 42°C (b) 25°C (c) -196°C (this happens to be the temperature of boiling liquid nitrogen) 12. 15.5 liters 13. 636°C 14. 921 ml 15. 555 ft^3 (this question shows why underwater laboratories open to the sea are pressurized; it keeps the water out!) 16. P_{Ar}=420 torr, P_{He}=240 torr, P_T=660 torr 17. p_{N_2}=300 torr, P_{O_2}= (680-300) = 380 torr 18. 741 torr 19. 29.4 psi 20. 525 ml O_2 21. 9.62 liters 22. (a) 82.1 ml atm/mole K (b) 62.4 liter torr/mole K 23. 0.123 mole 24. 0.208 atm 25. 28.0 g/mole 26. 20.5 liters 27. 1.25 g/liter 28. 58.0 g/mole, C_4H_{10} 29. The lighter (lower molecular weight) molecule diffuses faster. (a) H_2O (b) NH_3 (c) CO_2 30. 1.24 31. 1.004 32. 72 amu 33. b 34. d 35. see Figure 6.12 in the text 36. (a) CH_4 (b) H_2O (c) C_2H_5OH (d) CH_4

7 SOLIDS

Solids, in general, are characterized by highly ordered crystalline structures. In fact, it is this high degree of order that has enabled scientists to investigate their structures. Much of the knowledge that is possessed today about molecular structure, from simple salts such as sodium chloride to complex biomolecules like hemoglobin, has come from studies of crystalline solids.

There are three main topics developed in this chapter. One is the description of crystals in terms of a small number of repeating arrays of points called lattices. A second is the study of crystal structure by X-ray diffraction, and the third is the relationship between physical properties and the kinds of attractive forces present in solids.

7.1 CRYSTALLINE SOLIDS

Objectives

To learn what external features identify crystalline solids.

Review

Remember that crystals are characterized by flat faces that intersect at certain characteristic angles. In crystals of sodium chloride, for instance, the faces always intersect at 90° angles. Amorphous solids do not exhibit this property. Broken glass usually has curved surfaces that intersect at random angles.

Self-Test (Fill in the blanks)

1. An example of an amorphous solid is ________________

2. If a block of NaCl is crushed, each tiny crystal will possess interfacial angles equal to ________ degrees.

New Terms

amorphous
crystalline
crystal face
interfacial angle

7.2 X-RAY DIFFRACTION

Objectives

To learn how the structure of a crystal is investigated by X-ray diffraction.

Review

When an X-ray beam is directed on a crystal the atoms composing the crystal scatter the beam in all directions. The X-rays emerging from the crystal are only in phase in certain directions, however, and an intense X-ray beam is observed to come out of the crystal only at certain angles with respect to the incoming beam. The key point in this section is that the distance of separation between planes of atoms in a crystal is related to the angle at which an X-ray beam is observed to be reflected.

The Bragg equation relates the angle of reflection (θ), the distance between planes of atoms (d) and the wavelength of the X-rays (λ). Using X-rays of known wavelength the distances between atoms in the crystal can be calculated. Bragg's equation is

$$n\lambda = 2d \sin\theta$$

where n is an integer (n = 1 or 2 or 3, etc.)

Self-Test

3. X-rays of wavelength 1.54 Å are reflected from layers of atoms in a crystal of potassium chloride at an angle of 14.1°. What is the distance between the layers? (Assume n = 1)

____

New Terms

diffraction

7.3 LATTICES

Objectives

To find a way of describing a repeating pattern, such as the structure of a crystal, in terms of a lattice.

Review

One of the main themes throughout this section (and the entire chapter) is that it is possible to describe the structures of millions of different crystals in terms of a very small set of only 14 three-dimensional lattices.

A lattice may be a one-, two- or three-dimensional geometric array of points. To describe a crystal, of course, a three-dimensional lattice is required. The properties of the entire lattice are embodied in the unit cell. The entire lattice can be constructed by moving the unit cell along its edges by distances equal to the lengths of the edges. The unit cell is characterized by the lengths of its edges, $\underline{a}$, $\underline{b}$ and $\underline{c}$, and the angles opposite them, α, β and γ, respectively (see Figure 7.8 in the text).

There are four types of unit cells:

(1) primitive: lattice points only at the corners.

(2) body-centered: lattice points at the corners and in the center of the cell.

(3) face-centered: lattice points at the corners and in the center of each face.

(4) end-centered: lattice points at each corner and a pair of lattice points in the center of one pair of opposite faces.

Remember that only 1/8 of a corner point lies within a given unit cell, 1/4 of a point along an edge lies in a given cell, 1/2 of a face-centered point lies within a given cell, and any point within the cell contributes one entire point to the cell.

Self-Test

4. Consider the unit cell drawn on the following page.

   (a) What type of unit cell is it? ______________________

(b) How many lattice points are within it? ____________

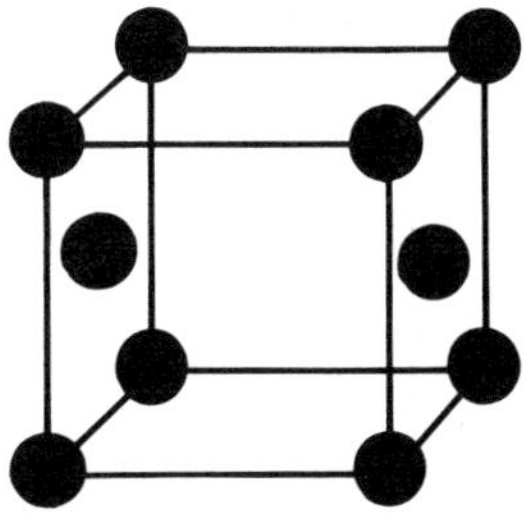

New Terms

lattice
space lattice
unit cell

7.4 AVOGADRO'S NUMBER

Objectives

To see how Avogadro's number can be calculated from the dimensions of a unit cell and a knowledge of crystal structure.

Review

Avogadro's number is so huge that it's very difficult to comprehend its magnitude. If you were to count one molecule per second, it would take approximately 191,000 billion centuries to count 6.02×10^{23} molecules! How can we possibly measure a number this large? The object of this section is to show one way of doing this, making use of the dimensions of the unit cell and a knowledge of crystal structure.

New Terms

none

7.5 ATOMIC AND IONIC RADII

Objectives

To see how crystal structure and unit cell dimensions can be used to determine the sizes of atoms and ions.

Review

Using simple geometry, the dimension of the unit cell and the crystal structure of a solid, it is possible to calculate atomic and ionic radii. Use reasoning similar to that shown in the text to answer the following Self-Test question.

Self-Test

5. Potassium chloride crystallizes with a face-centered cubic structure. One face of a unit cell is shown below. The unit cell edge is 6.28 Å long and the radius of the Cl^- ion is 1.81 Å. What is the ionic radius of K^+?

____

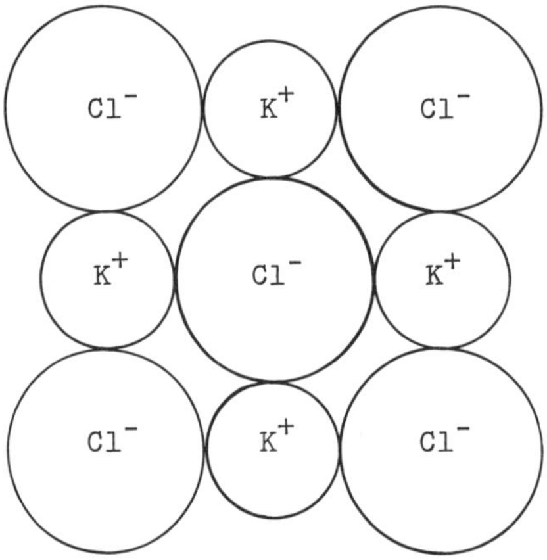

New Terms

none

7.6 THE FACE-CENTERED CUBIC LATTICE

Objectives

To see how the face-centered cubic lattice can be used to describe the structures of several different types of chemical compounds.

Review

In this section you are shown three different types of chemical structures, each possessing a face-centered cubic lattice.

(1) Rock salt structure (Figure 7.10 in the text) Each ion in the structure is surrounded octahedrally by ions of opposite charge. There is a 1:1 ratio of cations to anions.

(2) Zinc blende structure (Figure 7.11 in the text) Each ion is surrounded by four ions of opposite charge at the corners of a tetrahedron. There is a 1:1 ratio of cations to anions.

(3) Fluorite structure (Figure 7.12 in the text) Each ion is surrounded tetrahedrally by four ions of opposite charge. There is a 1:2 ratio of cations to anions. The antifluorite structure has the positions of cations and anions reversed.

Self-Test

6. Show that the zinc blende structure has a cation to anion ratio of 1:1.

7. What type of structures do the following substances have?

(a) ZnS ____________________

(b) CaO ____________________

(c) CaF_2 ____________________

(d) Na_2S ____________________

New Terms

rock salt structure	fluorite structure
zinc blende structure	antifluorite structure

7.7 CLOSEST-PACKED STRUCTURES

Objectives

To study how atoms can be packed in the most efficient manner.

Review

There are two types of closest packing:

(1) cubic closest-packing (ccp) This gives rise to a face-centered cubic lattice.

(2) hexagonal closest-packing (hcp) This gives an hexagonal lattice.

In both structures there are layers of tightly packed spheres stacked one above the other. The two structures differ only in the way the third layer is stacked above the first. In the hcp the third layer is directly above the first to give an ABABAB... repeating pattern. In the ccp the third layer atoms lie above depressions in the first to give an ABCABCABC... pattern.

In both structures there are empty spaces surrounded by atoms. Some of these are surrounded by four atoms and are called tetrahedral sites. Others are surrounded by six atoms at the corners of an octahedron and are called octahedral sites. Your text describes how some of the structures discussed in the last section can be viewed as cubic closest-packing of anions with cations in tetrahedral or octahedral sites.

Self-Test

8. Sodim chloride can be viewed as a ccp arrangement of Cl^- ions with Na^+ ions in octahedral sites. The fluorite structure can be viewed as a ccp arrangement of Ca^{2+} ions with F^- ions in tetrahedral sites.

   (a) How many Na^+ ions (open circles in Figure 7.10) are in the unit cell of NaCl (this is the number of octahedral sites in one unit cell)? ________

   (b) How many F^- ions (open circles in Figure 7.12) are in the unit cell of CaF_2 (this is the number of tetrahedral sites in one unit cell)? ________

   (c) What is the ratio of tetrahedral to octahedral sites? ________

New Terms

closest-packed structures
cubic closest-packing
hexagonal closest-packing
tetrahedral site
octahedral site

7.8 TYPES OF CRYSTALS

Objectives

To understand how the properties of a crystalline substance depend on the kinds of species that occupy sites in the lattice and on the nature of the attractive forces between them.

Review

The key points of this section are summarized in Table 7.3 in the text. Study this table well before answering the following Self-Test.

Self-Test

9. What crystal type is observed for

(a) NaCl ______________

(b) SO_2 ______________

(c) Ni ______________

(d) $MgCl_2$ ______________

(e) SiC ______________

10. CO_2 forms soft crystals (dry ice) that sublime (evaporate) at -78°C. SiO_2, on the other hand, forms hard high melting crystals (sand). What crystal type does each of these form?

____

11. UF_6 forms soft crystals that melt at 64.5°C. What is the probable type of crystal formed by UF_6? ______________

12. An element, formerly called columbium, melts at 2468°C, is soft and shiny, and conducts electricity. What is this type of crystal?

____

New Terms

molecular crystal
ionic crystal
covalent crystal
metallic crystal

7.9 BAND THEORY OF SOLIDS

Objectives

To understand how metals conduct electricity. You should also learn what distinguishes an insulator from a conductor and why some substances behave as semiconductors.

Review

In a solid, atomic orbitals of the same or similar energies on all of the atoms in the crystal combine to form energy bands. The band formed from the valence shell orbitals is termed the valence band. All bands below the valence band are completely occupied by electrons, which are held tightly to their respective atoms. A band that is empty or partially filled is called a conduction band. The conduction band extends continuously throughout the solid and electrons in the conduction band move freely from atom to atom.

A conductor is a substance that has either a partially filled conduction band (like Na) or a conduction band that overlaps the valence band (like Mg). In an insulator there's a large energy gap between a filled valence band and the empty conduction band, and it is difficult to raise an electron into the conduction band. A semiconductor has a small gap between the valence band and conduction band and thermal energy is sufficient to raise some electrons to the conduction band. Increasing the temperature increases the population of the conduction band and increases the conductivity.

New Terms

band theory
valence band
conduction band
insulator
semiconductor

7.10 DEFECTS IN CRYSTALS

Objectives

To see the types of imperfections that occur in crystals and to see how they affect the properties of solids.

Review

In real crystals we do not find perfect packing. The imperfections that occur are termed lattice defects. A point defect results when only a few atoms are involved in the deviation from perfection.

Frenkel defects are caused by cations being moved from their normal lattice sites to positions between layers (called interstitial positions).

Schottky defects occur when cation and anion sites are left vacant.

Defects also occur when an impurity is incorporated into a crystal lattice. This is particularly important in semiconductors. If the impurity has fewer electrons than atoms of the host lattice, a p-type semiconductor occurs in which charge is transferred by the movement of positively charged "holes".

When the impurity has more electrons than atoms of the host lattice, an n-type semiconductor is produced in which charge is transported by the extra electrons supplied by the guest atoms.

Self-Test

13. What type of semiconductor occurs if a small quantity of

    (a) Ga is added to pure As? ____________

    (b) Se is added to pure As? ____________

    (c) P is added to pure Si? ____________

New Terms

lattice defect
point defect
Frenkel defect
Schottky defect
nonstoichiometric compound
interstitial
F-center
p-type semiconductor
n-type semiconductor

ANSWERS TO SELF-TEST QUESTIONS

1. glass 2. 90° 3. 3.16 Å 4. (a) end-centered (b) 2 5. 1.33 Å
6. cation: 8 corners (1/8 atom/corner) +
6 faces (1/2 atom/face) = 1 + 3 = <u>4 atoms</u>
anion: 4 atoms internally = <u>4 atoms</u>
cation to anion ratio = 1:1
7. (a) zinc blende (b) rock salt (c) fluorite (d) antifluorite
8. (a) 4 (b) 8 (c) 2:1 9. (a) ionic (b) molecular (c) metallic (d) ionic (e) covalent
10. CO_2-molecular, SiO_2-covalent
11. molecular
12. metallic (the element is now called niobium)
13. (a) p-type (b) n-type (c) n-type

8 LIQUIDS AND CHANGES OF STATE

After having discussed solids and gases in the last two chapters, we now turn our attention to the third state of matter, liquids. In this chapter are treated a number of properties that can be used to describe the liquid state. In addition, this chapter also examines the transitions between the different states (solid ⟶ liquid, liquid ⟶ gas, etc.).

8.1 GENERAL PROPERTIES OF LIQUIDS

Objectives

To review some of the general properties characteristic of liquids in general.

Review

In a sense, the liquid state results as a compromise between the strong attractive forces that hold particles in a well-ordered crystal and the violent molecular motion that tends to produce the highly disordered gaseous state. The following is a summary of some of the general properties of liquids.

A liquid is characterized by a constant volume and a shape that conforms to the contours of the container in which it's placed. Liquids are virtually incompressible and undergo only small changes in volume upon changes in temperature or pressure. Molecules diffuse very slowly through liquids, although much more rapidly than they do through solids. Liquids exhibit a property called surface tension, which tends to minimize their surface area. This causes liquid droplets to have a spherical shape, since for a given volume a sphere has the smallest surface area. Finally, liquids evaporate with the absorption of heat.

Self-Test

1. (Multiple choice) The reason molecules diffuse more slowly in liquids than in gases is

   (a) the molecules move more slowly in a liquid than in a gas

   (b) the strong attractive forces in a liquid hold the molecules in place

   (c) the molecules move slowly because the liquid cannot expand easily

   (d) the molecules are constantly colliding with others, thereby interfering with their movement

   ____

New Terms

surface tension
mean free path

8.2 HEAT OF VAPORIZATION

Objectives

To examine the energy changes that accompany the evaporation of a liquid.

Review

The molar heat of vaporization is the energy required to convert one mole of liquid to one mole of vapor. Quantitatively, it is the difference between the heat content of the vapor and the heat content of the liquid.

$$\Delta H_{vap} = H_{vap} - H_{liq}$$

Remember that neither H_{vap} nor H_{liq} can actually be measured; only their difference can be.

The heat of vaporization is useful because it provides a direct measure of the strengths of the attractive forces that exist between the molecules in the liquid.

Variations in ΔH_{vap} show that among hydrocarbons the attractive forces (London forces) increase with chain length. London forces also increase with molecular size. In this section we see that hydrogen bonding is important for HF, H_2O and NH_3.

Self-Test

2. Using the data in Table 8.1, arrange the following compounds in order of increasing strengths of intermolecular attractive forces: C_2H_6, HCl, H_2S, HF, SiH_4, NH_3. ____________________

3. Without referring to Table 8.1, choose the compound in each of the following pairs with the stronger intermolecular attractive forces.

   (a) PH_3 or AsH_3 ________

   (b) SiH_4 or CH_4 ________

   (c) H_2O or H_2S ________

New Terms

molar heat of vaporization
polarizability

8.3 VAPOR PRESSURE

Objectives

To understand why the vapor pressure of a liquid depends only on the temperature. You should understand the concept of dynamic equilibrium and you should learn how the effects of outside influences on an equilibrium can be predicted by application of Le Chatelier's principle. In particular, you should learn how temperature affects the vapor pressure of a liquid. You should be able to calculate ΔH_{vap} from vapor pressures at two different temperatures.

Review

A dynamic equilibrium exists when two opposing processes occur at the same speed.

The vapor pressure of a liquid is the pressure exerted by its vapor in dynamic equilibrium with the liquid. In this case molecules are evaporating from the liquid into a closed container at the same rate that molecules are returning to the liquid.

Various factors can influence the position of equilibrium, that is, the relative amounts of reactants and products in a chemical system or, in this case, the relative amounts of liquid and vapor. Le Chatelier's principle states that when a system at equilibrium is disturbed (so as to destroy the equilibrium) the system readjusts in a way that minimizes, or counteracts, the stress placed upon it. If the pressure on the system is increased by a decrease in volume, the system will respond (if it can) in a way that tends to reduce the pressure. If heat is added to a system, the system responds by undergoing a change that absorbs heat. In each case the system changes in a way that tends to absorb the stress placed on it. You will encounter Le Chatelier's principle again in Chapters 12 and 14. Learning to apply it now will make things easier for you later on.

The vapor pressure of a liquid increases with temperature. A graph of vapor pressure versus temperature gives a vapor pressure curve, illustrated in Figure 8.7 in the text. The vapor pressure curve ends at the critical temperature, the temperature above which a gas can no longer be condensed to liquid by the application of pressure. At the critical temperature a gas can be liquefied by application of the critical pressure.

The Clausius-Clapeyron equation (Equation 8.2 on Page 234) relates ΔH_{vap} to the vapor pressure at two temperatures.

Example 8.1

Mercury has a vapor pressure of 1.2×10^{-3} torr at $20^{\circ}C$. Its heat of vaporization is 59.4 kJ/mole. Calculate the vapor pressure of mercury at $30^{\circ}C$.

Solution

To solve the problem we use the Clausius-Clapeyron equation. First, let's tabulate the data.

	1	2
P	1.2×10^{-3} torr	?
T	293 K	303 K

Substituting (using R = 8.31 J/mole K and $\Delta H_{vap} = 5.94 \times 10^4$ J/mole)

$$\log\left(\frac{P_1}{P_2}\right) = \frac{-(5.94 \times 10^4 \text{ J/mole})}{2.303\ (8.31 \text{ J/mole K})}\left(\frac{1}{293 \text{ K}} - \frac{1}{303 \text{ K}}\right)$$

$$\log\left(\frac{P_1}{P_2}\right) = -0.350$$

Taking the antilog,

$$\frac{P_1}{P_2} = 0.447$$

Solving for P_2 and substituting for P_1 gives

$$P_2 = \frac{1.2 \times 10^{-3} \text{ torr}}{0.447}$$

$$= 2.7 \times 10^{-3} \text{ torr}$$

Self-Test

4. Consider the process, vapor $\rightleftharpoons$ liquid + heat

   (a) A decrease in pressure will increase the amount of ______________

   (b) An increase in temperature will decrease the amount of __________

5. (Multiple choice) The vapor pressure of a liquid increases with increasing temperature primarily because as the temperature rises,

   (a) the molecules of the vapor move more rapidly
   (b) a greater fraction of molecules can escape the liquid
   (c) the attractive forces between the molecules in the vapor decrease
   (d) the rate of return to the liquid increases

   ____

6. Toluene (used to make TNT) has a vapor pressure of 10 torr at 6.4°C and a vapor pressure of 40 torr at 31.8°C. Calculate ΔH_{vap} for toluene.

   ____

New Terms

dynamic equilibrium
equilibrium vapor pressure
Le Chatelier's principle
vapor pressure curve
critical temperature
critical pressure
Clausius-Clapeyron equation

8.4 BOILING POINT

Objectives

To define more precisely the term, boiling point, and to understand why boiling point changes with pressure. You should also learn how boiling point provides a measure of the strengths of the intermolecular attractive forces in a liquid.

Review

The boiling point is the temperature at which the vapor pressure of the liquid equals the prevailing atmospheric pressure. The normal boiling point (standard boiling point) is the temperature at which the vapor pressure equals 760 torr.

The boiling point provides an indication of the strengths of the attractive forces between liquid molecules. If the attractive forces are high, the vapor pressure at a given temperature is low because only a small fraction of molecules can escape the liquid. On the other hand, when weak attractive forces are present, a large fraction can escape and the vapor pressure is high. Liquids that have low vapor pressures at a given temperature have high boiling points, while those with high vapor pressures have low boiling points.

The abnormally high boiling points of NH_3, H_2O and HF provide evidence for hydrogen bonding in these substances.

Self-Test

7. Why does HF have a lower boiling point than H_2O even though it forms stronger hydrogen bonds? ____________________

____

8. What effect does an increase in pressure have on the boiling point of a liquid? ______________________________

9. What is inside the bubbles in boiling water? ____________

New Terms

boiling point
normal boiling point

8.5 FREEZING POINT

Objectives

To define freezing point in terms of dynamic equilibrium and to consider the energy changes that take place upon freezing and melting.

Review

At the freezing point of a liquid there is a dynamic equilibrium between molecules in the solid and liquid. Molecules leave the solid and enter the liquid at the same rate that molecules leave the liquid and attach themselves to the solid. The energy that must be removed from one mole of liquid to convert it to solid is called the molar heat of crystallization. This is equal in magnitude, but opposite in sign, to the molar heat of fusion - the energy needed to melt one mole of solid. Fusion means the same as melting (Can you guess how an electrical fuse works?). Remember that ΔH_{fus} is always much less than ΔH_{vap}.

Self-Test

10. Why doesn't the value of ΔH_{fus} give a direct measure of the strengths of the attractive forces in the solid? ______________________

____

11. What difference is there between the freezing point of a liquid and the melting point of a solid? ______________________

New Terms

molar heat of crystallization
molar heat of fusion

8.6 HEATING AND COOLING CURVES: CHANGES OF STATE

Objectives

To observe what changes take place when heat is gradually added to a solid, or gradually removed from a gas. You should learn the kinds of energy changes that take place along the various segments of a heating or cooling curve. Learn the definition of supercooling.

Review

On those portions of a heating or cooling curve where the temperature is changing (the slanted line segments in Figures 8.11 and 8.12) the kinetic energy of the molecules is changing. The horizontal segments, where the temperature remains constant, correspond to changes in potential energy during a phase change (gas $\leftrightarrow$ liquid, liquid $\leftrightarrow$ vapor).

Supercooling is a phenomenon that occurs when a liquid is cooled so rapidly that its temperature drops below the ordinary freezing point before the molecules have an opportunity to assemble themselves into the proper arrangement to form a crystal. Amorphous solids such as glass are actually supercooled liquids.

Self-Test

12. Refer to the cooling curve which follows to answer the following:

(a) Which line segment corresponds to the conversion of vapor to liquid? __________

(b) Which line segments correspond to changes in kinetic energy? __________

(c) Which line segments correspond to changes in potential energy? __________

(d) Use a dotted line to indicate the effect of supercooling.

(e) What is the boiling point of the substance? __________

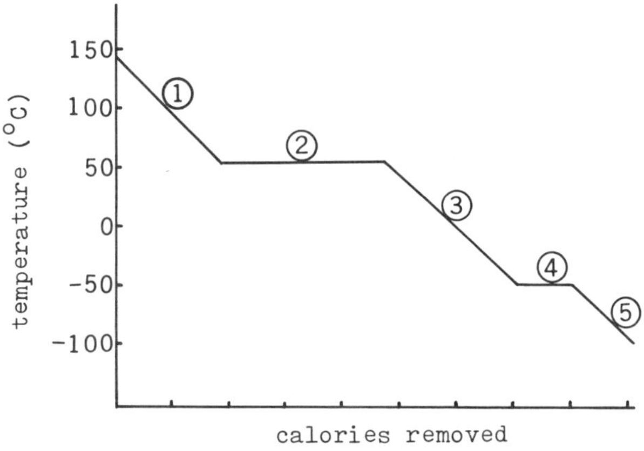

New Terms

heating curve
cooling curve
supercooling

8.7 VAPOR PRESSURE OF SOLIDS

Objectives

To learn how solids, like liquids, are able to evaporate by a process known as sublimation.

Review

Sublimation is the direct conversion of solid to vapor without passing through the liquid state. If the solid is placed in a closed container, the vapor can come to equilibrium with the solid. The pressure exerted by the vapor is called the vapor pressure of the solid. It too rises with increasing temperature.

Self-Test

13. The heat of sublimation, ΔH_{subl}, is the energy required to convert one mole of solid directly to one mole of vapor. The value of ΔH_{subl} is always greater than ΔH_{vap}. Why is this so? ______________________

____

14. Which substance would have a higher vapor pressure at -90°C, solid H_2O or solid CO_2? Why? ______________________

New Terms

sublimation
equilibrium vapor pressure of a solid

8.8 PHASE DIAGRAMS

Objectives

To learn how a phase diagram can be used to define the limits of temperature and pressure over which the different states of a substance can exist.

Review

The type of phase diagram discussed in this section contains three lines that define pressures and temperatures at which equilibria can exist between two phases. The solid-vapor equilibrium line is the vapor pressure curve for a solid. The liquid-vapor line, which terminates at the critical temperature and pressure, is the vapor pressure curve for the liquid. The solid-liquid line gives the melting point at different pressures. These lines serve as boundaries to temperature/pressure regions where only one phase can exist. Review Figure 8.15 and learn which regions of the diagram correspond to solid, liquid, and vapor. Remember that the three equilibrium lines intersect at the triple point, the temperature and pressure at which all three states can coexist in dynamic equilibrium.

The relationships between temperature, pressure and phase changes are covered in detail in Figures 8.16 to 8.18. Review this material too so that you can follow the changes that take place moving either horizontally or vertically on the phase diagram.

In most substances the solid-liquid line slants to the right. This is because in most cases the solid is more dense than the liquid. Application of pressure on a liquid (moving up vertically at constant temperature) converts it to the more dense (more compact) solid phase. You should be able to make this prediction on the basis of Le Chatelier's principle.

Self-Test

15. Sketch a phase diagram for a substance whose triple point occurs at $-10^{o}C$, 25 torr and whose normal melting and boiling points are $-5^{o}C$ and $120^{o}C$, respectively. Identify the solid-liquid (S-L), liquid-vapor (L-V) and solid-vapor (S-V) lines. Indicate the solid, liquid, and vapor regions.

New Terms

triple point
phase diagram

ANSWERS TO SELF-TEST QUESTIONS

1. d 2. $SiH_4 < C_2H_6 < HCl < H_2S < NH_3 < HF$ (arranged in order of increasing ΔH_{vap}) 3. (a) AsH_3 (b) SiH_4 (c) H_2O 4. (a) vapor (b) liquid 5. b 6. 38.7 kJ/mole 7. Because HF can form only two hydrogen bonds while H_2O can form four. Four weaker bonds turn out to be stronger than two strong bonds. 8. Increasing pressure raises the boiling point. 9. steam 10. It gives a measure of the difference between the strengths of attractive forces in two phases that each contain relatively strong attractive forces.

11. no difference - they are the same 12. (a) 2 (b) 1, 3, 5 (c) 2, 4 (d) see Figure 8.13 in the text (e) 50°C 13. The strengths of the attractive forces are greater in the solid than in the liquid. Therefore, more energy must be added to convert the solid to a gas than to change the liquid to a gas. 14. CO_2, because it has weaker attractive forces (CO_2 is nonpolar, H_2O is polar and exhibits hydrogen bonding).

15.

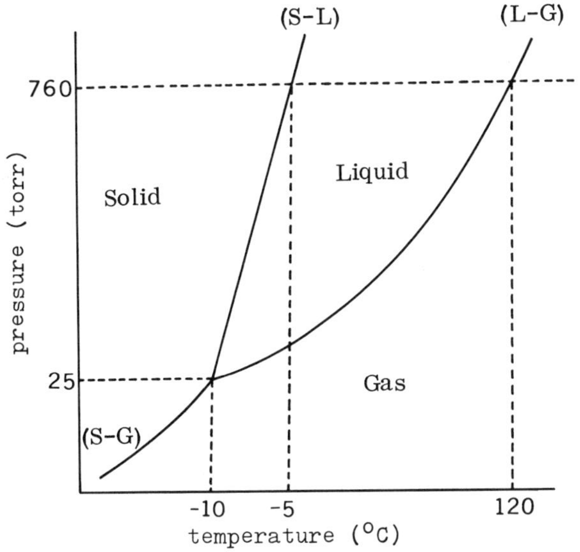

9 PROPERTIES OF SOLUTIONS

In Chapter 5 solutions were discussed in terms of their usefulness as a medium for carrying out chemical reactions. The focus of this chapter is on the physical properties of solutions (boiling point, freezing point, vapor pressure, etc.) as they are affected by the presence of a solute.

9.1 TYPES OF SOLUTIONS

Objectives

To review the different kinds of solutions that can be formed.

Review

Solutions can be formed between: gas-gas (gaseous solutions); liquid-solid (liquid solutions); solid-gas, solid-liquid, solid-solid (solid solutions).

Solid solutions can be of two types, substitutional and interstitial. Review the meanings of these.

Self-Test

1. What kind of solutions do the following represent?
   (a) black coffee (without sugar) __________
   (b) brass __________
   (c) a carbonated beverage __________
   (d) auto exhaust __________

(e) a martini (without the olive) ____________________

New Terms

substitutional solid solution
interstitial solid solution

9.2 CONCENTRATION UNITS

Objectives

To define additional concentration units that are useful in treating the physical properties of solutions. You should learn to convert between the different units.

Review

Be sure you know the definitions of the following units. The most common reason that students have difficulty handling concentration units is because they fail to learn the definitions.

mole fraction $$X_A = \frac{n_A}{n_A + n_B + n_C + \ldots}$$

n_A, n_B, n_C, etc. are the number of moles of each component of the solution.

weight fraction $$w_A = \frac{\text{weight of A}}{\text{total weight of solution}}$$

molarity $$M = \frac{\text{moles of solute}}{\text{liters of solution}}$$

normality $$N = \frac{\text{equivalents of solute}}{\text{liters of solution}}$$

molality $$m = \frac{\text{moles of solute}}{\text{kilograms of solvent}}$$

(Be careful to clearly distinguish between molarity and molality; they sound and look alike but are defined quite differently.)

Conversions among molality, mole fraction, and weight fraction are straightforward and require only molecular weights of solute and solvent (see Example 9.1 in the text). To convert any of these to molarity (or vice versa) requires the density of the solution, as shown in Example 9.3 and the additional example below.

Example 9.1

An aqueous solution of $CuSO_4$, having a mole fraction of $CuSO_4$ equal to 0.0127, has a density of 1.106 g/ml. What is the molarity of the $CuSO_4$?

Solution

We begin by assuming that we have a total of 1 mole of solute and solvent. Then there are

0.0127 mole $CuSO_4$
0.9873 mole H_2O

To obtain the volume of the solution, which we need to compute the molarity, we first must calculate the total mass of the solution and then apply the density.

$$0.0127 \text{ mole } CuSO_4 \left(\frac{159.5 \text{ g } CuSO_4}{1 \text{ mole } CuSO_4}\right) = 2.03 \text{ g } CuSO_4$$

$$0.9873 \text{ mole } H_2O \left(\frac{18.02 \text{ g } H_2O}{1 \text{ mole } H_2O}\right) = 17.79 \text{ g } H_2O$$

$$\text{Total weight of solution} = 19.82 \text{ g}$$

$$\text{volume of solution} = 19.82 \text{ g} \times \frac{1 \text{ ml}}{1.106 \text{ g}} = 17.92 \text{ ml}$$

We now have both the number of moles of $CuSO_4$ and the volume of the solution. The molarity is:

$$\text{molarity} = \frac{0.0127 \text{ mole } CuSO_4}{0.01792 \text{ liter soln.}}$$

$$= 0.709 \text{ M}$$

Self-Test

2. Calculate (a) the molality, (b) mole fraction of solute, and (c) weight fraction of solute in a solution prepared by dissolving 100 g of paraffin (molecular weight 370) in 500 g of benzene, C_6H_6. This solution could be used as a paint remover. ____________________

3. A solution of NaBr in water has a weight fraction of NaBr equal to 0.40. What is the (a) mole fraction, (b) mole percent, and (c) molality?

4. The density of the solution in Question 3 is 1.414 g/ml. What is the molarity of the NaBr?

New Terms

molality
weight fraction
volumetric flask
mole fraction
mole percent

9.3 THE SOLUTION PROCESS

Objectives

To learn what factors control the solubility of substances in liquid solvents.

Review

The key point in this section is that in order for substances to be appreciably soluble in each other, they must possess similar intermolecular attractive forces. Particles that attract each other very strongly tend to congregate and separate from those to which they are weakly attracted.

Remember that when a solute particle is placed in solution it becomes solvated, that is, surrounded by solvent molecules to which it is attracted. When the solvent is water, the term hydration is used.

Self-Test

5. How does hydration of ions help keep them in solution?
6. What does the term "like dissolves like" mean on a molecular level?

New Terms

hydration
solvation
miscible

9.4 HEATS OF SOLUTION

Objectives

To study the energy changes that occur when a solution is formed.

Review

The heat of solution is the energy absorbed or liberated when a solution is formed. In this section you saw that it is possible to divide the total energy change into various contributions, some of which are endothermic and some of which are exothermic.

For liquid solutions formed from a solvent (A) and solute (B), an ideal solution results when the A-B attractions are the same as the A-A and B-B attractions. For an ideal solution, $\Delta H_{soln} = 0$. When the A-B attractions are greater than the A-A and B-B attractions, $\Delta H_{soln} < 0$ and the solution process is exothermic. When the A-B attractions are weaker, $\Delta H_{soln} > 0$ and the solution process is endothermic.

For solutions of solids in liquids the lattice energy (the energy required to separate the solute particles from a crystal) and hydration energy (or solvation energy - the energy released when the solute particle is placed into the solvent cage) must be considered. This is summarized in Figure 9.9 in the text.

Self-Test

7. When acetone (a component of nail polish remover) is dissolved in water, the resulting solution becomes warm. What conclusions can you draw about the relative strength of the attractive forces between acetone and water molecules?

8. Acetone and water (Question 7 above) are completely soluble in each other in all proportions. Does this mean they form ideal solutions?

9. For LiCl, $\Delta H_{soln} < 0$. What does this imply about the relative values of the hydration energy of the ions and the lattice energy of LiCl?

New Terms

heat of solution	lattice energy
ideal solution	hydration energy

9.5 SOLUBILITY AND TEMPERATURE

Objectives

To examine the factors that determine the effect of temperature on solubility.

Review

A rise in temperature increases solubility if the dissolving of additional solute is endothermic. A fairly good rule of thumb is that the solubility of most solids and liquids in a liquid solvent increases with increasing temperature. The solubility of gases, however, almost always decreases with increasing temperature.

Self-Test

10. Why do gases usually become less soluble in liquids as the temperature of the solution is raised?

New Terms

none

9.6 FRACTIONAL CRYSTALIZATION

Objectives

To learn how the temperature effect on solubility can be used to purify substances.

Review

A solid substance containing a soluble impurity is dissolved in a minimum of hot solvent. The solution is cooled and some of the pure desired solid crystallizes, leaving the impurity behind in the solution along with some of the desired material. Even though some of the desired substance is lost, that which is recovered is usually of much higher purity.

The text describes the type of calculation that can be performed to recover the maximum amount of pure solid. In a practical sense, this kind of computation is rarely done. There are several reasons for this: (1) often the nature of the impurity is unknown; (2) if the impurity is known, its solubility at various temperatures usually is not; (3) the amount of impurity generally is unknown; (4) it's usually simpler and faster to simply dissolve the solid in hot solvent, cool the solution, and recover as much solid as possible.

Self-Test

11. A solid is known to contain 80 g of $NaNO_3$ and 5 g of NaCl. The solubility of NaCl is 36 g/100 g H_2O at 0^oC and 40 g/100 g H_2O at 100^oC. The solubility of $NaNO_3$ is 73 g/100 g H_2O at 0^oC and 180 g/100 g H_2O at 100^oC.

    (a) What is the minimum amount of boiling water (100^oC) necessary to dissolve all of the solid? __________

    (b) If the solution is cooled to 0^oC, how much $NaNO_3$ will separate as pure solid? __________

New Terms

fractional crystallization

9.7 THE EFFECT OF PRESSURE ON SOLUBILITY

Objectives

To learn how, and under what circumstances, pressure influences solubility.

Review

Pressure has virtually no effect on the solubility of solids or liquids in liquid solvents. The solubilities of gases, however, are very markedly affected by pressure changes. This can be predicted by Le Chatelier's principle since an increase in pressure favors a decrease in the number of moles of gas (this would tend to bring the pressure back down). Check with your instructor whether he expects you to treat the effect of pressure on the solubility of a gas quantitatively. If so, review the material below on Henry's law.

Henry's law relates the concentration of dissolved gas to its partial pressure over the solution.

$$C_g = k_g p_g$$

where k_g is the Henry's law constant.

Self-Test

12. Why are the solubilities of solids in liquids virtually unaffected by pressure?

13. Calculate the solubility of a gas, X, in water at 20^oC if its partial pressure is 720 torr ($k = 3.5 \times 10^{-3}$ g/liter torr at 20^oC). __________

New Terms

Henry's law

9.8 VAPOR PRESSURES OF SOLUTIONS

Objectives

To learn how the vapor pressure of a solution depends on the relative amounts of solute and solvent.

Review

When a nonvolatile, nondissociating solute is dissolved in a solvent, the vapor pressure of the solvent is diminished because a portion of the surface becomes occupied by molecules unable to enter the vapor phase. Raoult's law relates the vapor pressure to the mole fraction of the solvent in the solution,

$$P_{solution} = X_{solvent} P^o_{solvent}$$

where $P^o_{solvent}$ is the vapor pressure of the pure solvent.

When two volatile liquids are mixed, the vapor above the solution contains molecules of both. The vapor pressure of the solution is the sum of the partial pressures of each substance. The partial pressures are also determined by Raoult's law. For some substance, A, its partial pressure is

$$p_A = X_A P^o_A$$

Deviations from Raoult's law occur when the solution is nonideal (recall the definition of an ideal solution in Section 9.4). When ΔH_{soln} is negative, meaning heat is evolved as the solution is formed, the actual partial pressure of each component is less than that calculated from Raoult's law. The reason for this is that the solute and solvent are held more tightly in the solution than in either pure substance. These extra strong solute-solvent attractive forces are responsible for both a negative ΔH_{soln} and negative deviations from Raoult's law. Positive deviations occur when ΔH_{soln} is positive. In this case the A-B attractive forces are less than either A-A or B-B attractions. Molecules are held less tightly than predicted for an ideal solution and the vapor pressure is greater than that calculated from Raoult's law.

Self-Test

14. The vapor pressure of water at 100°C is 760 torr. What is the vapor pressure of a solution of 200 g of sugar, $C_{12}H_{22}O_{11}$, in 1000 g of H_2O at this same temperature? Will the solution boil at 100°C under an atmospheric pressure of 760 torr?

____

15. At 85°C, ethylene bromide ($C_2H_4Br_2$) has a vapor pressure of 170 torr and propylene bromide ($C_3H_6Br_2$) has a vapor pressure of 127 torr. These substances form very nearly an ideal solution. What is is the vapor pressure of a solution containing 100 g of each?

____

New Terms

Raoult's law

9.9 FRACTIONAL DISTILLATION

Objectives

To see how a distillation process can often be used to separate mixtures of volatile liquids.

Review

Remember that at the boiling point the sum of the partial pressures of the components above a mixture equals the atmospheric pressure.

$$P_{atm} = p_A + p_B = P_{Total}$$

The partial pressures, p_A and p_B, in turn, are found by Raoult's law.

$$p_A = X_A P_A^o$$

$$p_B = X_B P_B^o$$

Remember that with Raoult's law X_A and X_B are the mole fractions of A and B in the liquid.

In the vapor the mole fraction is found from Dalton's law. For example,

$$p_A = X_A P_{Total}$$

which gives

$$X_A = \frac{p_A}{P_{Total}} = \frac{p_A}{p_A + p_B}$$

Be careful not to confuse the mole fraction in the liquid with the mole fraction in the vapor.

When a liquid mixture boils, the vapor always has more of the more volatile component than does the liquid. Remember this because it can help you see when you've made a mistake in a calculation.

A boiling point curve is shown in Figure 9.18 in the text. Remember that the upper curve gives the composition of the vapor; the lower curve gives the composition of the liquid. The compositions of vapor and liquid in equilibrium are connected by a tie line. Review how repeated boiling and condensation of the vapor gradually gives a liquid richer in the more volatile component.

Mixtures having large deviations from Raoult's law form azeotropes. These mixtures have either maxima or minima in their boiling point curves. They can only be separated into one pure component plus the liquid mixture having the composition at the maximum or minimum of the curve.

Self-Test

16. Two substances, A and B, have vapor pressures at 85°C of 800 torr and 300 torr, respectively. What will be the composition of a mixture of these substances that boils at 85°C under 1 atm pressure? (Hint - use Raoult's law and remember that $X_B = 1 - X_A$)

____

New Terms

fractional distillation
boiling point diagram
tie line
azeotrope

9.10 COLLIGATIVE PROPERTIES OF SOLUTIONS

Objectives

To see how the vapor pressure lowering of a solution by a nonvolatile solute causes a boiling point elevation and freezing point depression. You should learn how this phenomenon permits the determination of molecular weights. In addition, you should learn how the dissociation of an electrolyte produces abnormally large changes in boiling and freezing points.

Review

You should examine Figure 9.20 in the text to be sure you understand why a nonvolatile solute raises the boiling point and lowers the freezing point.

Remember that

$$\Delta T_b = K_b m$$

and

$$\Delta T_f = K_f m$$

where m is the molality - the number of moles of solute particles per 1000 g (1 kilogram) of solvent. The specific values of K_b and K_f depend on the solvent (Table 9.5).

These relationships are useful in two ways. Knowing m and K_b or K_f, you can calculate the change in boiling and freezing points. This might be important, for example, if you wanted to know the properties of an antifreeze solution. The important application to chemistry is in the determination of molecular weights. Here we measure ΔT and knowing K we can calculate the molality. From a knowledge of the weight of solute and solvent we can obtain a relationship between weight and number of moles from which the molecular weight can easily be gotten.

Example 9.2

5.48 g of a solid are dissolved in 200 g of water to give a solution having a freezing point of $-0.850^{\circ}C$. What is the molecular weight of the substance?

Solution

From the freezing point depression, $0.850^{\circ}C$, we can get the molality.

$$m = \frac{\Delta T}{K} = \frac{0.850^\circ C}{1.86^\circ C/\text{molal}} = 0.457 \text{ molal}$$

This gives a ratio,

$$\frac{0.457 \text{ mole solid}}{1.00 \text{ kg } H_2O} \qquad (1)$$

Next we calculate the ratio of mass of solid to kilograms of water.

$$\text{ratio} = \frac{5.48 \text{ g solid}}{0.200 \text{ kg } H_2O}$$

Dividing numerator and denominator by 0.200 we get,

$$\text{ratio} = \frac{27.4 \text{ g solid}}{1.00 \text{ kg } H_2O} \qquad (2)$$

Ratio (1) must equal ratio (2) because we are dealing with the same solution. Since their denominators are the same, their numerators must be equal. The next step then is to equate numerators.

$$0.457 \text{ mole solid} = 27.4 \text{ g solid}$$

Finally, divide through by 0.457 to get the weight of one mole.

$$1 \text{ mole solid} = 60.0 \text{ g}$$

Electrolytes dissociate to produce more moles of particles than moles of solute. One mole of NaCl produces 2 moles of particles. The freezing point and boiling point changes are larger than they would be if no dissociation occurred. To obtain the ΔT, multiply the ΔT calculated using the molal concentration of salt by the number of ions produced when one formula unit dissociates. For example, for a 1.00 m solution of $CaCl_2$ we would calculate $\Delta T = 1.86^\circ C$. Since $CaCl_2$ produces three ions per $CaCl_2$ formula unit, the actual ΔT is three times as large as we originally calculated.

$$\Delta T_{\text{actual}} = 3(\Delta T_{\text{calculated}})$$

$$\Delta T_{\text{actual}} = 3(1.86^\circ C) = 5.58^\circ C$$

Self-Test

17. Calculate the boiling point elevation and the actual boiling point of a solution of 35.0 g of a solute having a molecular weight of 210 in 450 g of benzene. Use the data in Table 9.5.

____

18. What is the molecular weight of an unknown substance if a solution of 0.00213 g X in 0.100 g of camphor has a freezing point of 174.5°C? Use the data in Table 9.5.

____

19. What value of ΔT_f do you expect for 0.100 m solutions of:

(a) NaCl __________ (b) $Al_2(SO_4)_3$ __________

20. Calculate the freezing point depression produced by a solution of 0.100 g of a compound having a molecular weight of 10,000 dissolved in 100 g of water.

____

New Terms

colligative properties
molal boiling point elevation constant
molal freezing point depression constant
association
dimer

9.11 OSMOTIC PRESSURE

Objectives

To see how osmosis provides a means of determining very large molecular weights.

Review

If you worked through the solution to Question 19 above, you see that when the molecular weight is large, ΔT_f is very small and virtually impossible to measure. Measurements of osmotic pressure provide means of calculating very high molecular weights because the osmotic pressure produced by even very dilute solutions is measurable.

The van't Hoff equation is easy to remember - it looks like the ideal gas law.

$$\pi V = nRT$$

If you know π, V, R, and T, you can calculate the number of moles of solute in the solution.

Example 9.3

A 0.010-g sample of starch in 5.0 ml of water at 25°C produces a solution having an osmotic pressure of 2.3 torr. What is the average molecular weight of the starch molecules?

Solution

First, let's express π in atm.

$$\pi = 2.3 \text{ torr} \left(\frac{1 \text{ atm}}{760 \text{ torr}}\right) = 3.0 \times 10^{-3} \text{ atm}$$

We also have

$$V = 0.005 \text{ liter}$$
$$R = 0.0821 \text{ liter atm/mole K}$$
$$T = 298 \text{ K}$$

Solving the van't Hoff equation for n gives

$$n = \frac{\pi V}{RT}$$

and substituting,

$$n = \frac{(3.0 \times 10^{-3} \text{ atm})(0.005 \text{ liter})}{(0.0821 \text{ liter atm/mole K})(298 \text{ K})}$$

$$n = 6.1 \times 10^{-7} \text{ mole}$$

From the quantity of starch placed in the solution we have

$$6.1 \times 10^{-7} \text{ mole} = 0.010 \text{ g}$$

$$1 \text{ mole} = 1.6 \times 10^{4} \text{ g} = 16,000 \text{ g}$$

The molecular weight is 16,000.

Self-Test

21. A solution of a water-soluble polymer (0.2 g) in 10 ml of water at 20°C has an osmotic pressure of 0.10 torr. What is the molecular weight of the polymer? __________

New Terms

osmotic pressure
isotonic

9.12 INTERIONIC ATTRACTIONS

Objectives

To see that in solutions of electrolytes ions are not totally independent of each other.

Review

As the concentration of an electrolyte increases, its ions influence each other to a greater extent and are less independent. One way that this can happen is by the formation of ion pairs, groups of oppositely charged ions. As the concentration of ions increases, there is a greater chance for ions of opposite charge to encounter one another and an equilibrium concentration of ion pairs can be created. This in effect reduces the number of independent particles that are available to alter the properties of the solution.

The van't Hoff factor, i, is the ratio of the observed freezing point depression (or boiling point elevation) to the freezing point depression (or boiling point elevation) that the substance would exhibit if it were a non-electrolyte.

Self-Test

22. What is the limiting i factor (at infinite dilution) for the following:

(a) $MgSO_4$ __________ (c) $Al_2(SO_4)_3$ __________

(b) K_2SO_4 __________

New Terms

van't Hoff i factor

ANSWERS TO SELF-TEST QUESTIONS

1. (a) liquid-solid (b) solid-solid (c) liquid-gas (d) gas-gas (e) liquid-liquid 2. (a) 0.541 m (b) 0.04 (c) 0.166 3. (a) 0.104 (b) 10.4 (c) 6.48 m 4. 5.50 M 5. The polar water molecules help shield ions of opposite charge from each other. 6. Substances tend to be soluble in each other only if they have about equal intermolecular attractive forces. 7. Acetone attracts water more than acetone attracts acetone or water attracts water. 8. No. In Question 7 you saw that $\Delta H_{soln} < 0$. For an ideal solution $\Delta H_{soln} = 0$. 9. The hydration energies of the ions must be greater than

the lattice energy. 10. Because ΔH_{soln} for a gas is usually negative (exothermic). Increasing the temperature requires adding heat. Le Chatelier's principle predicts that an endothermic change should occur which requires that gas leave the liquid. 11. (a) 44.4 g H_2O (b) 48 g. Note that all of the NaCl is soluble in 44.4 g H_2O. 12. Because they are incompressible. 13. 2.52 g/liter 14. p_{H_2O} = 752 torr. No, the vapor pressure is less than atmospheric pressure. 15. 149.3 torr

16. Solution

$$p_A + p_B = 760$$

$$X_A P_A^o + X_B P_B^o = 760$$

$$X_A P_A^o + (1 - X_A)P_B^o = 760$$

$$X_A(800) + (1 - X_A)(300) = 760$$

$$800X_A + 300 - 300X_A = 760$$

$$500X_A = 760 - 300 = 460$$

$$X_A = 460/500$$

$$X_A = 0.92$$

$$X_B = 1 - X_A = 0.08$$

17. $\Delta T_b = 0.94$, $T_b = 80.1 + 0.94 = 81.0^oC$ 18. 188 g/mole
19. (a) 0.372^oC (b) 0.930^oC 20. 0.000186^oC
21. 3.7×10^6 g/mole 22. (a) 2 (b) 3 (c) 5

10 CHEMICAL THERMODYNAMICS

As pointed out in the text, Chapters 10 and 11 deal with the two factors that control whether or not products of a chemical reaction will form. Thermodynamics controls the feasibility of a reaction in the sense that it determines how much product can be formed; kinetics (Chapter 11) controls how fast the products are formed.

Thermodynamics deals with energy changes and is applied to physical as well as chemical changes. You will see that a number of thermodynamic principles are developed here using physical systems as examples, followed by the extension of the principles to chemical systems. The applications of thermodynamics range from simple chemical reactions to complex reactions in living organisms.

10.1 SOME COMMONLY USED TERMS

Objectives

To become familiar with some of the terminology to be used in later discussions.

Review

This section defines in a precise way some terms that have been used rather loosely before. It also introduces some new terms. Be sure of their meaning before moving on.

New Terms

system
surroundings
adiabatic
isothermal
state
state function
equation of state
heat capacity
specific heat
molar heat capacity

10.2 THE FIRST LAW OF THERMODYNAMICS

Objectives

To see how the law of conservation of energy applies to heat transfer and the performing of work.

Review

The first law is simply a restatement of the law of conservation of energy.

$$\Delta E = q - w$$

Remember the following:

$$\Delta E = E_{final} - E_{initial}$$

Also, q is energy (heat) added to the system.

w is energy removed from the system when the system performs work.

ΔE depends only on the initial and final states; E is a state function.

The entire discussion on the isothermal (constant temperature) expansion of an ideal gas was designed to show you that the magnitudes of q and w depend on the way the expansion is carried out - hence q and w are not state functions. Since ΔE is the same regardless of the path between initial and final states, E is a state function.

Remember that one way a system can perform work is to expand against an externally applied pressure. Under a constant pressure,

$$w = P\Delta V$$

Self-Test

1. A compressed gas in a cylinder pushes back a piston against a constant opposing force of 8.0 atm. The initial and final volumes of the gas are 25 ml and 600 ml. How much work, expressed in calories, is done by the gas during the expansion? ____________

   Express this answer in joules. ____________

New Terms

first law of thermodynamics
internal energy

10.3 REVERSIBLE AND IRREVERSIBLE PROCESSES

Objectives

To define what is meant by a reversible process and to see that the maximum amount of work is derived from a reversible change.

Review

A reversible change is one that takes place in an infinite number of steps, each of which takes place with the opposing force just barely below the driving force for the process. In the reversible expansion of a gas, for instance, the external pressure is initially high and equal to the internal pressure exerted by the gas. As the gas expands the external pressure is dropped at the same rate as the internal pressure drops so that the driving and restraining forces are essentially balanced throughout the entire expansion.

For the reversible expansion of a gas at constant temperature, the maximum work can be calculated from the equation,

$$w_{max} = 2.303 \, nRT \log \frac{V_f}{V_i}$$

where R is the gas constant and T is the absolute temperature and n is the number of moles of gas.

Self-Test

2. Calculate the work done by a gas as it expands from an initial 4 liters at 20 atm to 20 liters at 4 atm

   (a) by a one-step process against an opposing pressure of 4 atm. __________

   (b) by a two-step process in which the opposing pressure in the first step is 10 atm and in the second step is 4 atm. __________

   (c) by a reversible process at 25°C (Hint: first calculate the number of moles of gas.) __________

New Terms

reversible process

10.4 HEATS OF REACTION: THERMOCHEMISTRY

Objectives

To express energy changes in chemical reactions in terms of thermodynamic quantities.

Review

For changes at constant volume (and temperature), the heat of reaction is equal to ΔE.

$$\Delta E = q_v$$

For changes at constant pressure (and temperature), the heat of reaction is ΔH.

$$\Delta H = q_p$$

H is the enthalpy (also called heat content) and, like E, is a state function. The enthalpy is defined as

$$H = E + PV$$

and at constant pressure

$$\Delta H = \Delta E + P\,\Delta V$$

A calorimeter is a device used to measure heats of reaction. A bomb calorimeter has a constant volume. When reactions take place in it, the heat liberated is absorbed by the calorimeter and its temperature increases. From the temperature change and a knowledge of the heat capacity of the calorimeter, the amount of heat evolved in the reaction can be calculated (Example 10.1 in the text). The heat of reaction at constant pressure (ΔH) is measured in a similar way, but the contents of the calorimeter are kept at constant pressure.

Usually ΔH and ΔE are computed on a per mole basis to make them intensive properties, rather than extensive ones. For a reaction at constant pressure, the difference between ΔH and ΔE depends on the size of the volume change that accompanies the reaction. For reactions involving only liquids and solids, $\Delta H \approx \Delta E$. When gases are consumed or produced, the PV work is given by

$$P\,\Delta V = \Delta nRT$$

where Δn is the change in the number of moles of gas. Review Example 10.2 in the text.

Self-Test

3. One slice of bread plus sufficient oxygen for complete combustion are placed in a bomb calorimeter having a heat capacity of 36,500 cal/°C. The initial temperature of the calorimeter was 25.00°C. After the combustion was completed, the temperature rose to 26.64°C. How many nutritional Calories (1 Calorie, written with a capital C, equals 1000 calories, or 1 kcal) are contained in one slice of bread?

____

4. When 0.50 mole of methane, CH_4 (natural gas), is oxidized by oxygen to produce 0.50 mole of CO_2 and 1.0 mole of water vapor, 401 kJ of heat energy are evolved. The balanced equation for the reaction is:

$$CH_4(g) + 2\,O_2(g) \longrightarrow CO_2(g) + 2H_2O(g)$$

(a) What is ΔH in the units kJ/mole CH_4?

____

(b) What is ΔE at 25°C (expressed in the same units)?

____

5. Methanol may someday replace gasoline as a fuel in automobiles. It burns according to the equation,

$$2CH_3OH(l) + 3\,O_2(g) \longrightarrow 2CO_2(g) + 4H_2O(g)$$

Oxidation of one mole of CH_3OH at 25°C and constant pressure liberates 1280 kJ. What is ΔE for this reaction expressed in kJ/mole?

____

New Terms

enthalpy
calorimeter

10.5 HESS' LAW OF HEAT SUMMATION

Objectives

To use the fact that H is a state function in calculating values of ΔH. You should learn how to combine ΔH values when chemical equations are added or subtracted to produce new equations as well as the definition of heat of formation.

Review

A chemical equation written to show the energy change that takes place is called a thermochemical equation. Thermochemical equations are always interpreted on a mole basis; they therefore may be written with fractional coefficients. Remember to always indicate the physical state (solid, liquid or gas) of the substances written in a thermochemical equation.

Hess' law says, in effect, that the ΔH for some net reaction is the sum of all of the ΔH's for steps along the way. When thermochemical equations are added together to obtain some final equation, the ΔH for the final equation is the sum of the ΔH's of the thermochemical equations that were combined.

Example 10.1

Add the thermochemical equations below to obtain the value of ΔH for the reaction,

$$2Na_2O_2(s) + 4HCl(g) \longrightarrow 4NaCl(s) + 2H_2O(l) + O_2(g)$$

Equations:

$$2Na_2O_2(s) + 2H_2O(l) \longrightarrow 4NaOH(s) + O_2(g) \qquad \Delta H = -30.2 \text{ kcal}$$

$$NaOH(s) + HCl(g) \longrightarrow NaCl(s) + H_2O(l) \qquad \Delta H = -42.8 \text{ kcal}$$

Solution

Inspection of the two equations that are to be combined reveals that we must have 4NaOH on the left to cancel with the 4NaOH on the right when we add the equations. This means the second equation, and its ΔH, must be multiplied by 4 before adding it to the first.

$2Na_2O_2(s) + 2H_2O(l) \longrightarrow 4NaOH(s) + O_2(g)$ $\Delta H = -30.2$ kcal

$4NaOH(s) + 4HCl(g) \longrightarrow 4NaCl(s) + 4H_2O(l)$ $\Delta H = 4(-42.8$ kcal$)$
$= -171.2$ kcal

Adding the equations gives

$$2Na_2O_2(s) + 2H_2O(l) + 4NaOH(s) + 4HCl(g) \longrightarrow 4NaOH(s) + O_2(g) + 4NaCl(s) + 4H_2O(l)$$

Adding their ΔH values,

$$\Delta H = (-30.2 \text{ kcal}) + (-171.2 \text{ kcal}) = -201.4 \text{ kcal}$$

Remember that in problems of this type, if you have to reverse an equation in order to get cancellation of unwanted formulas, you must also reverse the sign of ΔH.

The heat of formation, ΔH_f, is the enthalpy change that occurs when a substance is formed from its elements. Hess' law can be restated in terms of heats of formation as:

$$\Delta H = (\text{sum of } \Delta H_f \text{ of the products}) - (\text{sum of } \Delta H_f \text{ of the reactants})$$

Self-Test

6. Use the equations
$CaO(s) + SO_3(g) \longrightarrow CaSO_4(s)$ $\Delta H = -95.9$ kcal

$Ca(OH)_2(s) \longrightarrow CaO(s) + H_2O(g)$ $\Delta H = +26.1$ kcal

to obtain the value of ΔH for the reaction,
$Ca(OH)_2(s) + SO_3(g) \longrightarrow CaSO_4(s) + H_2O(g)$ __________

7. Use the equations
$2C_2H_2(g) + 5\,O_2(g) \longrightarrow 4CO_2(g) + 2H_2O(g)$ $\Delta H = -2512$ kJ

$N_2(g) + 1/2\,O_2(g) \longrightarrow N_2O(g)$ $\Delta H = +104$ kJ

to obtain ΔH for the reaction,
$C_2H_2(g) + 5N_2O(g) \longrightarrow 2CO_2(g) + H_2O(g) + 5N_2(g)$

____

8. Use the equations
$2NO(g) + O_2(g) \longrightarrow 2NO_2(g)$ $\Delta H = -105$ kJ

$2N_2O(g) + 3\,O_2(g) \longrightarrow 4NO_2(g)$ $\Delta H = -55$ kJ

$NO_2(g) + SO_2(g) \longrightarrow NO(g) + SO_3(g)$ $\Delta H = -47$ kJ

to calculate ΔH for the reaction,

$2NO(g) + SO_2(g) \longrightarrow N_2O(g) + SO_3(g)$ ______________

New Terms

thermochemical equation
Hess' law of heat summation
enthalpy diagram
heat of formation

10.6 STANDARD STATES

Objectives

To establish standard conditions for the comparison of heats of reaction. You should learn to apply Hess' law using standard heats of formation.

Review

Standard conditions are chosen to be 25°C (298 K) and 1 atm pressure. A substance in its natural state under these conditions is said to be in its standard state. Standard states are indicated by a superscript zero.

Standard heats of formation, such as those in Table 10.1 in the text, can be used to calculate standard heats of reaction following Hess' law.

$$\Delta H^o = (\text{sum of } \Delta H_f^o \text{ products}) - (\text{sum of } \Delta H_f^o \text{ reactants})$$

This is illustrated in Example 10.3 in the text. Remember that we always take ΔH_f^o for any pure element in its standard state to be equal to zero.

Sometimes standard heats of formation cannot be measured directly. In these cases a reaction is carried out in which the heat of reaction is measured and the heats of formation of all reactants and products, except the one compound in question, are known. The unknown ΔH_f^o can then be solved for. Example 10.5 in the text illustrates this method using a measured heat of combustion.

Self-Test

9. Use the data in Table 10.1 to calculate ΔH^o for the following reactions:

(a) $CH_4(g) + 4Cl_2(g) \longrightarrow CCl_4(l) + 4HCl(g)$ ______________

(b) $Fe_2O_3(s) + 3CO(g) \longrightarrow 3CO_2(g) + 2Fe(s)$ ______________

(c) $C_2H_5OH(l) + O_2(g) \longrightarrow CH_3COOH(l) + H_2O(g)$ _____

10. The heat of combustion of octane, $C_8H_{18}(l)$, a component of gasoline, to produce $CO_2(g)$ and $H_2O(g)$ is -1213.6 kcal/mole of C_8H_{18}. What is the value of ΔH_f of $C_8H_{18}(l)$? _____

11. From the data in Table 10.1 can you suggest why using N_2O instead of O_2 in a combustion reaction produces a higher flame temperature?

New Terms

standard state

10.7 BOND ENERGIES

Objectives

To see how thermodynamic data can be used to obtain information about the strengths of chemical bonds. You should learn how to use bond energies to obtain an estimate of the heat of formation of a compound.

Review

The bond energy is the energy necessary to break a bond to produce neutral fragments. The atomization energy, ΔH_{atom}, is the energy needed to break all of the bonds in a molecule to give neutral atoms.

In many cases the tabulated average bond energies are additive in the sense that they may be used to calculate an atomization energy of a molecule. This can be used as one part of an alternate path from the free elements in their standard states to the compound in question in its standard state.

separated neutral atoms $\longrightarrow$ gaseous compound

$\Delta H_2 = -\Delta H_{atom}$

$\Delta H_1 =$ sum of ΔH_f of gaseous atoms

$\Delta H_3 = \Delta H$ for conversion of gas to compound in its standard state

elements $\longrightarrow$ compound
(standard state) ΔH_f^o (standard state)

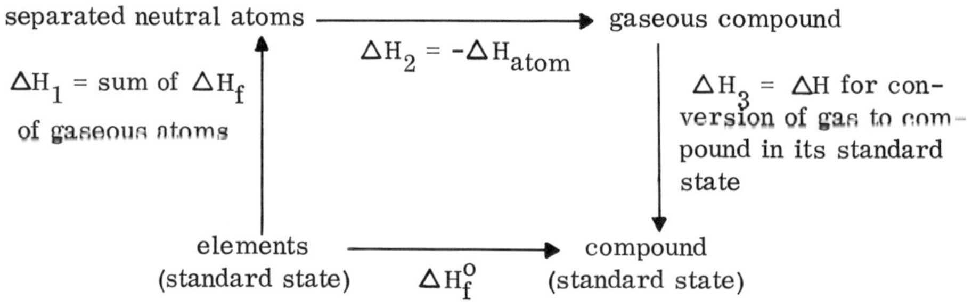

If the standard state of the compound is the gas, ΔH_3 can be ignored.

Example 10.2

Calculate the value of ΔH_f^o for dimethyl ether vapor,

$$H - \underset{\underset{H}{|}}{\overset{\overset{H}{|}}{C}} - O - \underset{\underset{H}{|}}{\overset{\overset{H}{|}}{C}} - H$$

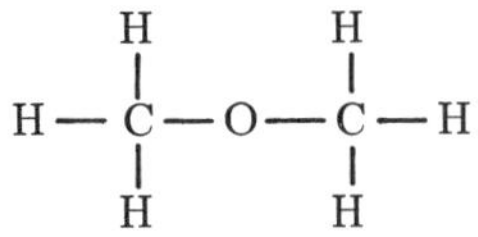

Solution

The reaction whose ΔH we wish to calculate is shown below with the alternate path indicated below it.

$$2C(s, \text{ graphite}) + 3H_2(g) + 1/2 \, O_2(g) \xrightarrow{\Delta H_f^o} C_2H_6O(g)$$

$$\downarrow 2\Delta H_f[C(g)] \quad \downarrow 6\Delta H_f[H(g)] \quad \downarrow \Delta H_f[O(g)] \qquad \uparrow -\Delta H_{atom}$$

$$2C(g) \quad + \quad 6H(g) \quad + \quad O(g)$$

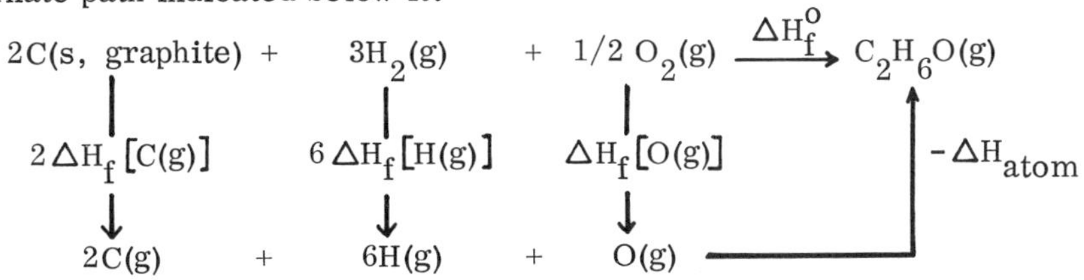

The minus sign appears before the ΔH_{atom} because in the direction of the arrow the process is the revers of atomization. When a reaction is reversed, remember that the sign of ΔH is reversed too.

The sum of all ΔH's along the lower path must equal the ΔH along the upper path (that is, ΔH_f^o).

$$\Delta H_f^o = 2\Delta H_f[C(g)] + 6\Delta H_f[H(g)] + \Delta H_f[O(g)] - \Delta H_{atom}$$

From Table 10.2 we can get the first three terms on the right.

$$\Delta H_f^o = 2(+715 \text{ kJ}) + 6(+218 \text{ kJ}) + (+249 \text{ kJ}) - \Delta H_{atom}$$

ΔH_{atom} is calculated from the number and kind of bonds in the molecule.

6(C – H) bonds	6(415)	= 2490 kJ
2(C – O) bonds	2(356)	= 712 kJ
	ΔH_{atom}	= 3202 kJ

Substituting,

$$\Delta H_f^o = 1430 \text{ kJ} + 1308 \text{ kJ} + 249 \text{ kJ} - 3202 \text{ kJ}$$

$$\Delta H_f^o = -215 \text{ kJ}$$

Since we are dealing with the formation of one mole of product,

$$\Delta H_f^o = -215 \text{ kJ/mole}$$

(The experimentally measured value is -185 kJ/mole.)

Self-Test

12. Calculate the atomization energy for acetic acid,

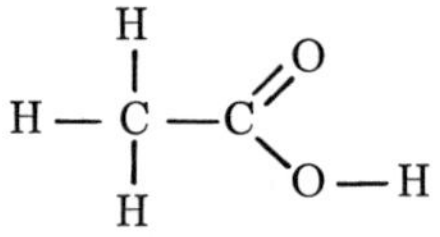

13. Use the data in Tables 10.2 and 10.3 to calculate ΔH_f^o for $C_2H_6(g)$. Compare your value with that found in Table 10.1. The structure of C_2H_6 is

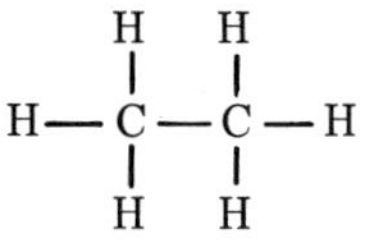

14. Use the data in Tables 10.2 and 10.3 to calculate ΔH_f^o for $C_6H_6(l)$. The heat of vaporization of C_6H_6 is 8.19 kcal/mole. The molecule exists as a resonance hybrid,

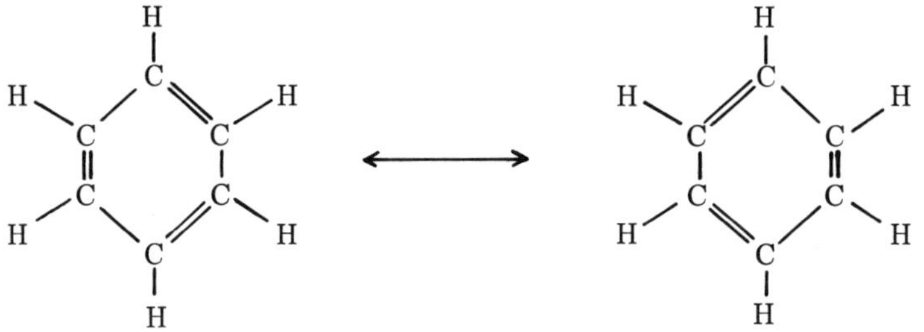

How does your calculated value compare with that found in Table 10.1?

New Terms

atomization energy

10.8 SPONTANEITY OF CHEMICAL REACTIONS

Objectives

To examine the factors that favor spontaneity in chemical and physical processes.

Review

The main point developed in this section is that there are two factors that influence spontaneity. A process tends to occur spontaneously (that is, without any outside help) if it takes place with a lowering of its energy. Therefore, exothermic reactions tend to occur spontaneously. It is also found that a physical or chemical change is favored when the system proceeds from a less probable state to one of greater probability. This corresponds to an increase in the degree of randomness, or disorder, in the arrangement of particles in the system.

Self-Test

15. You know that in tossing a coin there is an equal probability of it coming up either heads or tails. Suppose you had two coins, labeled A and B, and were to toss them. Make a table showing all of the possible combinations of heads and tails for the two coins. What is the probability that both coins would come up heads?

16. Do the same thing as you did in Question 15, but for 4 coins, labeled A, B, C, D. What is the probability of all four coins coming up heads when tossed? What is the probability of there being two heads and two tails?

New Terms

spontaneity

10.9 ENTROPY

Objectives

To describe a thermodynamic state function that can be associated with randomness in a chemical or physical system.

Review

The entropy, S, is a quantity which is a measure of the probability of a given state of a system. Since disorder or randomness is a more probable arrangement of particles than a highly ordered arrangement, entropy is said to increase as the disorder of the system increases.

The entropy change, ΔS, is given by

$$\Delta S = q_{rev}/T$$

where q_{rev} is the heat absorbed by a system if it were to go from one state to another by a reversible process and T is the absolute temperature at which the heat is added to the system.

New Terms

entropy

10.10 THE SECOND LAW OF THERMODYNAMICS

Objectives

To obtain an equation that incorporates the two factors that favor spontaneous changes (i.e., decrease in energy, increase in entropy).

Review

Chemical and physical changes are favored when accompanied by a decrease in energy. A process also tends to be spontaneous if it occurs with an increase in entropy.

The second law of thermodynamics states, in essence, that whenever a spontaneous process occurs, the entropy of the universe increases.

The Gibbs free energy is defined as

$$G = H - TS$$

At constant temperature and pressure,

$$\Delta G = \Delta H - T\Delta S$$

Remember that for a process to be spontaneous ΔG must be negative. In terms of the right side of the above equation, the difference, $\Delta H - T\Delta S$, must be negative. This section considers three possibilities:

(1) ΔH negative, ΔS positive: ΔG will be negative at all temperatures. The process will be spontaneous regardless of the temperature.

(2) ΔH positive, ΔS negative: ΔG will be positive at all temperatures. The process cannot be spontaneous at any temperature.

(3) ΔH and ΔS both positive or both negative: the sign of ΔG depends on the value of T. If ΔH and ΔS are both positive, for example, ΔG becomes the difference between two positive quantities (ΔH and $T\Delta S$). At high temperature $T\Delta S$ can be larger than ΔH; ΔG will be negative and the process will be spontaneous. At low temperature ΔH will be larger than $T\Delta S$; ΔG will be positive and the process won't be spontaneous.

Self-Test

17. What will be the sign of ΔG at high temperature if both ΔH and ΔS are negative? __________

18. The conversion of liquid CCl_4 to gaseous CCl_4 at 1 atm occurs with $\Delta H = +32.8$ kJ/mole and $\Delta S = +95.0$ J/mole K. Above what temperature should this process occur spontaneously? __________

New Terms

Gibbs free energy
second law of thermodynamics

10.11 FREE ENERGY AND USEFUL WORK

Objectives

To see how ΔG is related to the useful work that can be extracted from a system during a spontaneous change.

Review

The free energy change, ΔG, is equal to the maximum work that can be gotten during a spontaneous change. Remember that this maximum work can only be obtained if the change is reversible. All real systems undergo changes in an irreversible manner; therefore, the amount of work that can be extracted from a real system is somewhat less than the maximum predicted by ΔG.

New Terms

none

10.12 FREE ENERGY AND EQUILIBRIUM

Objectives

To see the relationship between free energy and equilibrium.

Review

The main point in this section is that $\Delta G = 0$ when a system is at equilibrium.

New Terms

none

10.13 STANDARD ENTROPIES AND FREE ENERGIES

Objectives

To establish standard entropies and free energies of substances. You should learn the third law of thermodynamics. You should also learn to calculate ΔG^o for reactions from ΔG_f^o. In addition, you should learn the qualitative relationship between ΔG^o and position of equilibrium.

Review

The third law of thermodynamics states that for any pure substance, S = 0 at 0 K. Absolute values of entropy can be obtained and some are given in Table 10.4. Values of ΔG^o can be gotten from calculated ΔH^o and ΔS^o, the latter obtained by suitably combining the S^o of products and reactants.

$$\Delta S^o = (\text{Sum } S^o \text{ products}) - (\text{Sum } S^o \text{ reactants})$$

Table 10.5 tabulates values of ΔG_f^o. These can be used in Hess' law type calculations to compute values of ΔG^o for reactions.

$$\Delta G^o = (\text{sum of } \Delta G_f^o \text{ products}) - (\text{sum of } \Delta G_f^o \text{ reactants})$$

One of the major points in this section is that ΔG^o is related to the position of equilibrium. Review the discussion centering on Figures 10.12 and 10.13. Notice that even with a ΔG^o that is positive some reaction occurs, because ΔG starts out negative heading in the direction of the products. Also note that the distinction is made here between ΔG^o and ΔG. ΔG^o is the difference between the free energy of the products in their standard states and the free energy of the reactants in their standard states. We are using ΔG, on the other hand, to stand for the way that the free energy is changing as we move along the free energy curve. Our earlier discussion about spontaneity applies in the sense that the free energy must be decreasing (ΔG negative) when the process is occurring spontaneously. Once the minimum is reached, equilibrium is established because the system cannot climb out of the free energy well. Therefore, ΔG^o serves as a guide to the feasibility of reactions.

Self-Test

19. Calculate values of ΔS_f^o for the following:

(a) $Al_2O_3(s)$ ____________________

(b) $C_3H_8(g)$ ____________________

(c) $PbSO_4(s)$ ____________________

20. The standard entropy of $CS_2(g)$ is 237.7 J/mole K. Use the data in Tables 10.1 and 10.4 to calculate ΔG_f^o of $CS_2(g)$. ____________________

21. Calculate ΔG^o for the following reactions:

(a) $H_2SO_4(l) + CaO(s) \longrightarrow CaSO_4(s) + H_2O(l)$ ____________________

(b) $Ag(s) + 2HNO_3(l) \longrightarrow AgNO_3(s) + NO_2(g) + H_2O(l)$

____

22. Sketch free energy curves for the reactions:

(a) $N_2(g) + 1/2\ O_2(g) \longrightarrow N_2O(g)$

(b) $H_2(g) + Cl_2(g) \longrightarrow 2HCl(g)$

23. Is the reaction, $N_2(g) + O_2(g) \longrightarrow 2NO(g)$, thermodynamically feasible at room temperature? Is it more or less feasible at 1000°C? (Assume ΔH^o and ΔS^o are essentially independent of temperature.)

____

New Terms

third law of thermodynamics
standard entropies
feasibility of reaction

10.14 APPLICATIONS OF THE PRINCIPLES OF THERMODYNAMICS

Objectives

To obtain a feel for the broad applicability of the principles of thermodynamics.

Review

Several examples of the influence of thermodynamics on our lives were discussed in this section in an effort to give you an appreciation of the scope and general applicability of the principles discussed in this chapter. We will have occasion to use these concepts in later chapters in discussions of equilibrium and the properties of the elements and their compounds.

New Terms

none

ANSWERS TO SELF-TEST QUESTIONS

1. 111 cal, 464 J 2. (a) 64 l-atm (b) 88 l-atm (c) 129 l-atm 3. 59.9 Calories 4. (a) 802 kJ/mole (b) 802 kJ/mole ($P\ \Delta V = 0$) 5. -1284 kJ/mole 6. -69.8 kcal 7. -1777 kJ 8. -177 kJ 9. (a) -102.6 kcal (429.1 kJ) (b) -6.6 kcal (-29.8 kJ) (c) -107.8 kcal (-451 kJ) 10. -59.4 kcal/mole 11. Because N_2O has a positive ΔH_f^o, when it serves as a reactant it tends to make ΔH more negative (since ΔH of reactants are subtracted from those of the products). 12. 749.6 kcal/mole (3136 kJ/mole) 13. calculated $\Delta H_f^o = -24.5$ kcal/mole (-100 kJ/mole) 14. calculated $\Delta H_f^o = +49.7$ kcal/mole (+209 kJ/mole), actual $\Delta H_f^o = +11.7$ kcal/mole (+49.0 kJ/mole). Note that the molecule is more stable than we expected. In general, molecules that exhibit resonance are more stable than one might expect.

15.

A	B
H	H
H	T
T	H
T	T

Probability of both heads = 1/4 = 0.25

16.

A	B	C	D	
H	H	H	H	*
H	T	H	H	
H	H	T	H	
H	H	H	T	
H	T	T	H	**
H	T	H	T	**
H	H	T	T	**
H	T	T	T	
T	H	H	H	
T	T	H	H	**
T	H	T	H	**
T	H	H	T	**
T	T	T	H	
T	T	H	T	
T	H	T	T	
T	T	T	T	

16 possible combinations

* = all heads

** = 2 heads, 2 tails

probability all heads = 1/16 = 0.0625

probability 2 heads, 2 tails = 6/16 = 0.375

17. ΔG will be positive 18. 345 K (72°C). When $\Delta G = 0$, $\Delta H = T\Delta S$, $T = \Delta H/\Delta S$. Above this T, ΔG will be negative. 19. (a) -74.85 cal/mole K (-313.1 J/mole K) (b) -64.41 cal/mole K (-269.6 J/mole K) (c) -85.6 cal/mole K (-358 J/mole K) 20. $\Delta G_f^o = \Delta H_f^o - (298\ K)\Delta S_f^o$; $\Delta G_f^o = +16.0$ kJ/mole 21. (a) -63.0 kcal (-262 kJ) (b) -13.8 kcal (-57.3 kJ)
22. (a)

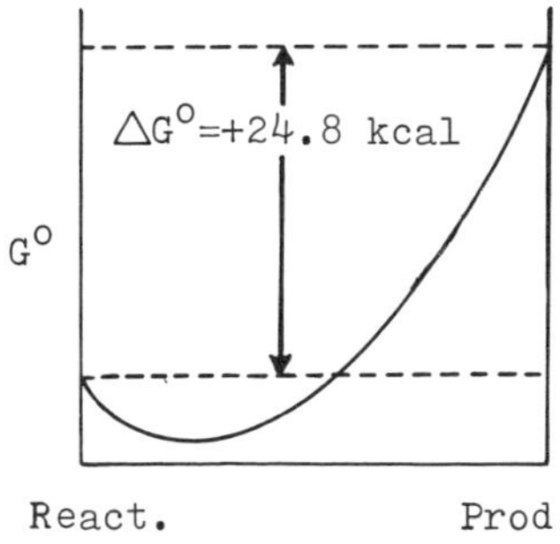

(b)

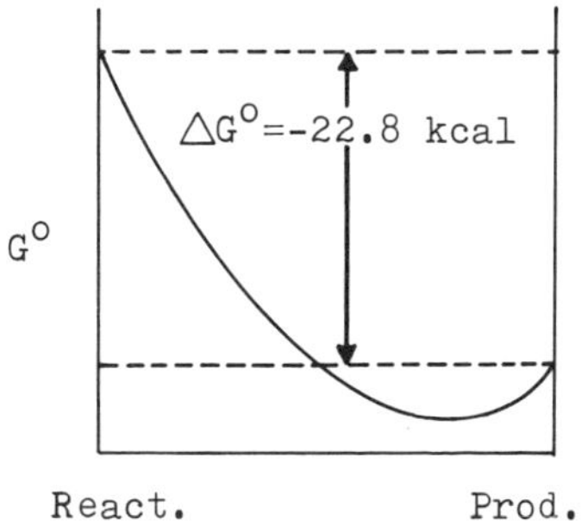

23. Since $\Delta G^o = +41.4$ kcal, the reaction is not feasible because very little NO will be produced. At 1000°C (1273 K),

$$\Delta G = \Delta H^o - (1273\ K)\Delta S^o$$
$$= +43.2\ kcal - (1273\ K)(+5.91\ cal/mole\ K)$$
$$= +43.2\ kcal - 7.5\ kcal = +35.7\ kcal$$

Since ΔG is smaller, the reaction will be a little more feasible at the higher temperature.

11 CHEMICAL KINETICS

This chapter concerns itself with the rates at which chemical reactions take place. This subject is important for several reasons. First, unless a reaction proceeds at a measurable rate, no products will be observed regardless of how thermodynamically favorable the reaction might be. Second, a study of reaction rates and the factors that influence them provides insight into the sequence of chemical steps that occurs to produce the overall net reaction. Third, a study of the effect of temperature on reaction rate gives information about energy changes that occur along the path from reactants to products.

11.1 REACTION RATES AND THEIR MEASUREMENT

Objectives

To see what is meant by reaction rate and to see how it is measured. You should also become aware of the units used to express reaction rates.

Review

The term "reaction rate" describes how fast the concentrations of reactants or products change with time and is usually expressed in the units mole/liter second. For most reactions the rate changes (decreases) as the reactants are consumed.

The rate of reaction can be obtained from a graph of the concentration of a reactant (or product) versus time (see Figure 11.2). The rate at some time, t, is obtained from the slope of the tangent to the concentration-time curve at time t. The procedure is described in Figure 11.2.

Remember that square brackets, [ ], denote molar concentration.

Self-Test

1. The rate of a reaction was found to be 3.0×10^{-4} mole/liter second. What would be the rate if it were expressed in the units, mole/liter minute?

____

2. From the concentration vs time curve below, estimate the rate of reaction at t = 10 sec.

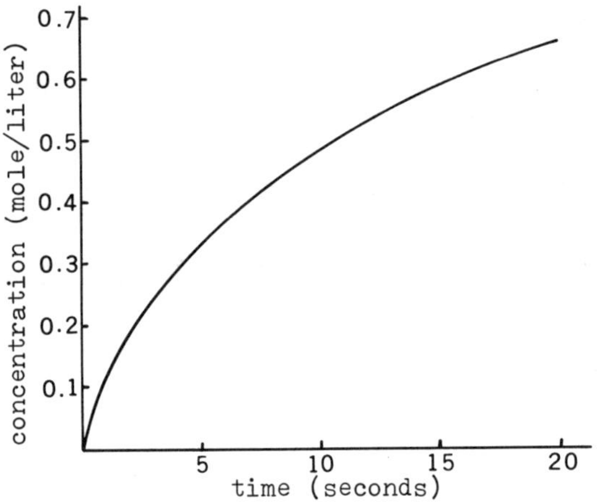

3. The sequence of chemical reactions that provides the net overall change in a chemical reaction is called ________________

4. List the four factors that affect the rate of a chemical reaction.

____

____

New Terms

mechanism
reaction rate

11.2 RATE LAWS

Objectives

To learn what is meant by the term, rate law, and to see how it can be obtained from experimental data.

Review

The rate law for a reaction such as

$$A + B \longrightarrow \text{products}$$

is

$$\text{Rate} = k[A]^x[B]^y$$

where k is the rate constant. The exponents, x and y, are called the order of the reaction with respect to A and B, respectively. The overall order is (x + y).

It is very important to remember that the actual values of x and y can only be obtained by experiment. These experiments involve observing what effect altering the concentrations of the reactants has on the rate of reaction.

For a reaction, $A \longrightarrow B$,

(1) if doubling the concentration of A doubles the rate, the exponent is 1.

$$\text{Rate} = k[A]^1$$

(2) if doubling the concentration of A quadruples the rate, the exponent is 2.

$$\text{Rate} = k[A]^2$$

Review Example 11.1 in the text to see how these rules are applied. Notice that once the exponents in the rate law have been established, the rate constant can be evaluated from any of the sets of data.

Self-Test

5. Below are some typical rate laws. What are the orders of the reactions and the orders with respect to each reactant?

   (a) $\text{Rate} = k[NO]^2[Br_2]$
   $2NO + Br_2 \longrightarrow 2NOBr$ ________________

   (b) $\text{Rate} = k[NO]^2[H_2]$
   $2NO + 2H_2 \longrightarrow N_2 + 2H_2O$ ________________

6. The following data were collected for the reaction,

$$A + 2B \longrightarrow C + D$$

concentrations A	B	rate (mole/liter sec)
0.10	0.10	2.0×10^{-4}
0.10	0.20	4.0×10^{-4}
0.20	0.20	1.6×10^{-3}
0.30	0.20	3.6×10^{-3}

(a) The rate law for the reaction is ____________

(b) The value of the rate constant is ____________

(c) The units of the rate constant are ____________

New Terms

order of reaction
rate law

11.3 COLLISION THEORY

Objectives

To obtain a theory that accounts quantitatively for the dependence of reaction rate on concentration.

Review

The basis for the collision theory is the notion that molecules must collide in order to react with one another. We can predict the rate law if we know what collisions take place during a reaction. Usually reactions occur in a series of steps before the ultimate products have been formed and we don't actually know what these steps are. In fact, one of the goals of kinetics is to give us a way of guessing intelligently at what these steps might be.

Remember, if we have the following collision processes, their rate laws are:

$A + B \longrightarrow$ products Rate = $k[A][B]$

$A + A \longrightarrow$ products

or $2A \longrightarrow$ products Rate = $k[A]^2$

New Terms

collision theory
bimolecular collision
termolecular collision

11.4 REACTION MECHANISM

Objectives

To see how collision theory helps us choose between alternate possible mechanisms for a reaction. You should learn the meaning of "rate determining step".

Review

The individual reactions that make up a mechanism are called elementary processes. The slowest step in the mechanism is called the rate determining step because the final products can't be formed any faster than the products of the slowest step.

Obtaining a satisfactory mechanism for a reaction is a very difficult task. A chemist must draw on all his experience and knowledge to arrive at a set of elementary processes that both make sense chemically and fit the experimentally determined rate law. With your limited chemical background you can't be expected to derive chemically reasonable mechanisms. However, within reason, you should be able to decide from a comparison of predicted and experimentally found rate laws whether a given mechanism is possible. You should also be able to decide which step in a mechanism must be the slow step in order to yield the correct rate law.

Remember that the predicted rate law should only include the reactants in the overall equation. Any intermediate products should not appear. In the very simple mechanisms that we are dealing with you can get the exponents in the predicted rate law by adding together all of the steps up to and including the rate determining step. The coefficients of the reactants at this point are the exponents in the predicted rate law. Try this with the mechanism proposed for the reaction between NO and H_2 at the beginning of the section in the text. The second reaction is the rate determining step.

Self-Test

7. Consider the following mechanisms:

$$\begin{array}{ll}(1) & 2A \longrightarrow Q \\ (2) & Q + B \longrightarrow C + D \\ (3) & B + D \longrightarrow 2M\end{array}$$

(a) What is the equation for the overall reaction?

____

(b) What is the rate law if step 1 is rate determining? __________

(c) What is the rate law if step 2 is rate determining? __________

8. The following mechanism was proposed to account for a chemical reaction. The rate law was found experimentally to be: Rate = $k[R]^2$

$$\begin{array}{l}2R \longrightarrow X + Y \\ X + Z \longrightarrow T + U \\ U + Z \longrightarrow P\end{array}$$

Which is the rate determining step? __________________

New Terms

elementary process
rate determining step

11.5 EFFECTIVE COLLISIONS

Objectives

To understand why reaction rates are nearly always much less than the rate of collision between molecules.

Review

The first thing to keep in mind here is that except for a very few cases, reactions proceed at a much slower rate than we would at first expect on the basis of the frequency of molecular collisions. Two factors tend to limit the number of collisions that are effective. One is the energy of the colliding molecules; the other is related to the orientation of the molecules when they collide.

New Terms

none

11.6 TRANSITION STATE THEORY

Objectives

To follow the energy changes that take place during an effective collision. Also, to see how the rate of reaction is affected by the energy required to produce an effective collision. You should become familiar with the potential energy diagram for a reaction.

Review

The minimum kinetic energy required between two colliding molecules in order to produce an effective collision is called the activation energy, E_a. Review the energy diagrams in Figures 11.4 to 11.6. You should be able to identify:

(1) the potential energy of the reactants
(2) the potential energy of the products
(3) the activation energy for the forward reaction
(4) the activation energy for the reverse reaction
(5) the heat of reaction
(6) whether the forward reaction is endo- or exothermic

The species that exists at the top of the potential energy diagram (i.e., the species that exists during an effective collision) is called the activated complex. The peak on the potential energy diagram is called the transition state. Transition state theory is concerned with the characteristics (geometry, energy) of the activated complex.

As a general rule, reactions having high activation energies tend to occur slowly, whereas fast reactions usually have low activation energies.

Self-Test

9. Identify the following by the numbers on the potential energy diagram which follows.

(a) _____ the potential energy of the products

(b) _____ the activation energy for the forward reaction

(c) _____ the heat of reaction

(d) _____ the potential energy of the reactants

(e) _____ the activation energy of the reverse reaction

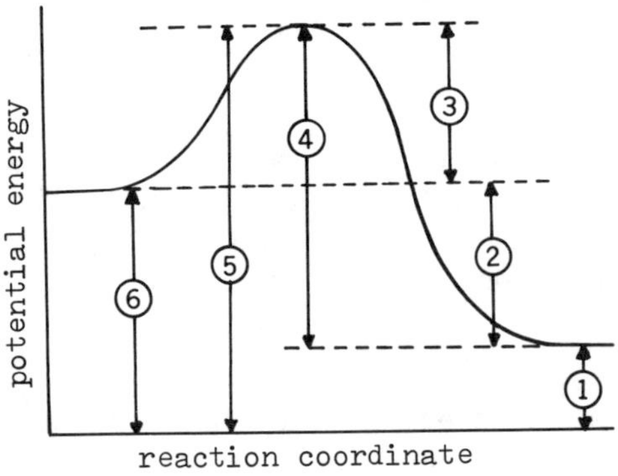

New Terms

activation energy
reaction coordinate
activated complex
transition state

11.7 EFFECT OF TEMPERATURE ON REACTION RATE

Objectives

To learn how and why temperature affects the rate of reaction. You should be able to calculate the activation energy from rate constants at two different temperatures.

Review

In very nearly every case, increasing the temperature increases the rate of reaction. This is because molecules move faster, on the average, at higher temperatures and more molecular collisions have the minimum kinetic energy to produce a net chemical change.

The Arrhenius equation is

$$k = Ae^{-E_a/RT}$$

From this can be derived the expression,

$$\ln \frac{k_1}{k_2} = \frac{E_a}{R}\left(\frac{1}{T_2} - \frac{1}{T_1}\right)$$

or, in terms of common logarithms,

$$\log \frac{k_1}{k_2} = \frac{E_a}{2.303\ R}\left(\frac{1}{T_2} - \frac{1}{T_1}\right)$$

If rate constants k_1 and k_2 are known at temperatures T_1 and T_2, the activation energy can be calculated.

Self-Test

10. As a rough rule of thumb, the rates of many reactions approximately double for every _________ ^oC rise in temperature.

11. The rate constant of a reaction at 15^oC is 1.3×10^{-5} liter/mole sec, while at 50^oC its rate constant is 8.0×10^{-3} mole/liter sec. What is E_a for the reaction?
____

12. The activation energy for a certain reaction was found to be 25.0 kcal/mole. At 25^oC the rate constant is 2×10^{-3} sec^{-1}. What is the rate constant at 50^oC?
____

New Terms

Arrhenius equation

11.8 CATALYSTS

Objectives

To define what a catalyst is and to understand how it functions in affecting the rate of reaction.

Review

A catalyst is a substance that alters the rate of a reaction by providing an alternate path (mechanism) from reactants to products that has a lower activation energy than the uncatalyzed reaction. It is important to remember that a catalyst functions by changing the mechanism.

A homogeneous catalyst exists in the same phase as the reactants. Biological enzymes are examples of homogeneous catalysts that promote specific biochemical reactions.

A heterogeneous catalyst exists as a separate phase from the reactants and products and appears to function by adsorbing reactant molecules on its surface where the reaction can somehow proceed more readily. Heterogeneous catalysts are widely used in industrial applications because they don't have to be separated later from the products of reaction as would homogeneous catalysts.

An inhibitor is a substance that becomes adsorbed on the surface of the catalyst and thereby decreases, or inhibits, its activity. The catalyst is then said to be poisoned.

Self-Test

13. How does a catalyst alter the activation energy of a reaction?

____

14. What type of catalyst is present in a catalytic muffler?

____

15. What type of catalyst is a biological enzyme?

____

New Terms

catalyst	heterogeneous catalyst
homogeneous catalyst	inhibitor

11.9 CHAIN REACTIONS

Objectives

To examine a type of mechanism that often occurs in systems with very complicated rate laws. You should learn that these reactions tend to propagate themselves once they have been started. You should learn the kinds of steps responsible for initiation, continuation, and termination of the chain.

Review

Chain reactions often involve free radicals - molecules or ions that contain unpaired electrons. These unpaired electrons tend to pair up with electrons in other molecules or ions, and free radicals are very reactive. The kinds of reactions that may be found in a chain mechanism are:

(1) initiation - this is the step that generates the first free radical.

(2) propagation - a product is formed plus another free radical. This continues the chain.

(3) inhibition - this is a step that slows down the rate of formation of products. In the mechanism in the text it removes some product but still generates a free radical so the chain can continue.

(4) termination - this is a step that removes free radicals and therefore interrupts the chain.

Self-Test

16. The reaction of methane with chlorine ($CH_4 + Cl_2 \longrightarrow CH_3Cl + HCl$) is believed to occur by the following mechanisms. Identify the nature of each of these reactions.

(a) $Cl_2 \xrightarrow{\text{light}} 2Cl\cdot$ ____________

(b) $Cl\cdot + CH_4 \longrightarrow HCl + CH_3\cdot$ ____________

(c) $CH_3\cdot + Cl_2 \longrightarrow CH_3Cl + Cl\cdot$ ____________

(d) $Cl\cdot + Cl\cdot \longrightarrow Cl_2$ ____________

(e) $CH_3\cdot + CH_3\cdot \longrightarrow C_2H_6$ ____________

(f) $CH_3\cdot + Cl\cdot \longrightarrow CH_3Cl$ ____________

ANSWERS TO SELF-TEST QUESTIONS

1. 1.8×10^{-2} mole/liter minute 2. 0.026 mole/liter sec 3. the mechanism 4. nature of the reactants, concentration of reactants, temperature, catalysts 5. (a) second order in NO, first order in Br_2, third order overall (b) second order in NO, first order in H_2, third order overall 6. (a) Rate = $k[A]^2[B]$ (b) k = 0.2 (c) $\text{liter}^2/\text{mole}^2$ sec 7. (a) $2A + 2B \longrightarrow C + 2M$ (b) Rate = $k[A]^2$ (c) Rate = $k[A]^2[B]$ 8. The first step 9. (a) 1 (b) 3 (c) 2 (d) 6 (e) 4 10. $10^\circ C$ 11. 34.0 kcal/mole (143 kJ/mole) 12. 5.2×10^{-2} sec^{-1} 13. a catalyst provides a different, low energy mechanism 14. heterogeneous catalyst 15. homogeneous catalyst 16. (a) initiation (b) propagation (c) propagation (d) termination (e) termination (f) termination

12 CHEMICAL EQUILIBRIUM

As a chemical reaction proceeds, the concentrations of the reactants decrease and the rate of the forward reaction decreases. At the same time, the concentrations of the products increase and the rate of the reverse reaction increases. Eventually both reactions occur at the same rate and dynamic equilibrium is achieved.

There is a simple relationship between the concentrations of reactants and products in an equilibrium system. This chapter deals with that relationship, how it can be understood from both the standpoint of kinetics and thermodynamics, and how it can be used in calculations relating equilibrium concentrations.

12.1 THE LAW OF MASS ACTION

Objectives

To establish the relationships between reactant and product concentrations in a chemical equilibrium. You should note that this section views the equilibrium law as a purely experimentally measurable phenomenon without attempting to present an explanation for it.

Review

Remember that the mass action expression for any chemical reaction can be written from the balanced equation. The concentrations of the products always appear in the numerator, raised to powers that are equal to their coefficients in the balanced equation. The concentrations of reactants are multiplied together in the denominator where they are raised to powers equal to the coefficients of the reactants in the balanced equation.

At equilibrium, at a given temperature, the mass action expression is always equal to the same number, the equilibrium constant. Remember, there are no restrictions on the individual equilibrium concentrations. The only requirement is that when they are substituted into the mass action expression, the resulting fraction always has the same value.

For gaseous reactions the equilibrium constant expression can be written in terms of concentration, thereby giving K_c, or in terms of partial pressures, giving K_p.

Self-Test

1. Write the mass action law giving both K_c and K_p for each of the following gaseous reactions:

(a) $PCl_5 \rightleftharpoons PCl_3 + Cl_2$

(b) $Br_2 + SO_2 + 2H_2O \rightleftharpoons H_2SO_4 + 2HBr$

(c) $6XeF_4 + 12H_2O \rightleftharpoons 2XeO_3 + 4Xe + O_2 + 24HF$

(d) $2NO_2 + F_2 \rightleftharpoons 2NO_2F$

2. Below are equilibrium concentrations of NO_2 and N_2O_4. The equilibrium equation is: $N_2O_4(g) \rightleftharpoons 2NO_2(g)$

Experiment	$[N_2O_4]$	$[NO_2]$
1	4.46×10^{-2} M	3.11 M
2	1.50×10^{-3} M	0.571 M
3	2.30×10^{-7} M	7.06×10^{-3} M
4	1 M	14.7 M

Show that they obey the mass action law. What is the value of the equilibrium constant?

____

New Terms

law of mass action
mass action expression
equilibrium constant

12.2 THE EQUILIBRIUM CONSTANT

Objectives

To see how the magnitude of K provides an immediate qualitative estimate of the extent to which a reaction proceeds toward completion.

Review

When K is large the reaction proceeds far toward completion; when K is small hardly any products are present at equilibrium.

Self-Test

3. Arrange the following reactions in order of increasing tendency to proceed toward completion:

(a) $CO(g) + Cl_2(g) \rightleftharpoons COCl_2(g)$ $K = 5 \times 10^9$

(b) $N_2O_4(g) \rightleftharpoons 2NO_2(g)$ $K = 217$

(c) $2SO_2(g) + O_2(g) \rightleftharpoons 2SO_3(g)$ $K = 8 \times 10^{25}$

(d) $2HCl(g) \rightleftharpoons H_2(g) + Cl_2(g)$ $K = 3.1 \times 10^{-17}$

____

New Terms

none

12.3 KINETICS AND EQUILIBRIUM

Objectives

To see how our understanding of chemical kinetics can be used to derive the law of mass action.

Review

The entire thrust of this section is to demonstrate that in an equilibrium, each step in a reaction mechanism becomes itself an equilibrium. The same mass action expression is obtained regardless of the mechanism of the reaction.

New Terms

none

12.4 THERMODYNAMICS AND CHEMICAL EQUILIBRIUM

Objectives

To relate, quantitatively, the standard free energy change for a reaction to the equilibrium constant. After completing this section you should be able to compute $\underline{K}$ from ΔG^o.

Review

The free energy change for a reaction is related to the mass action expression (Q) by the equation,

$$\Delta G = \Delta G^o + 2.303\ RT \log Q \qquad (12.2 \text{ in text})$$

Important equations to remember are

$$\Delta G^o = -2.303\ RT \log \underline{K}_p \qquad (12.4 \text{ in text})$$

for gaseous reactions, and

$$\Delta G^o = -2.303\ RT \log \underline{K}_c \qquad (12.5 \text{ in text})$$

for reactions in solution. The $\underline{K}$ that you compute using these equations is called the thermodynamic equilibrium constant.

Example 12.1

What is the thermodynamic equilibrium constant, $\underline{K}_p$, for the reaction, $2HBr(g) + Cl_2(g) \rightleftharpoons 2HCl(g) + Br_2(g)$, at 25°C?

Solution

First we calculate ΔG^o from the appropriate ΔG_f^o in Table 10.5.

$$\Delta G^o = 2\Delta G_f^o(HCl) - 2\Delta G_f^o(HBr)$$

$$\Delta G^o = 2\ \text{mole}(-22.8\ \text{kcal/mole}) - 2\ \text{mole}(-12.7\ \text{kcal/mole})$$

$$\Delta G^o = -20.2\ \text{kcal}$$

Next we solve Equation 12.4 in the text for log $\underline{K}_p$.

$$\log \underline{K}_p = \frac{-\Delta G^o}{2.303\ RT}$$

In these calculations, remember:

(1) use R = 1.987 cal/mole K if ΔG^o is in kcal; use R = 8.314 J/mole K if ΔG^o is in kJ.

(2) make sure you express ΔG^o in calories (or J), not kcal (or kJ).

(3) use the absolute temperature (298 K in this question).

Substituting,

$$\log \underline{K}_p = \frac{-(-20200)}{2.303(1.987)(2.98)} = 14.8$$

$$\underline{K}_p = 10^{14.8}$$

$$= 6 \times 10^{14}$$

Self-Test

4. Calculate the value of the thermodynamic equilibrium constant at 25°C for the reaction, $H_2(g) + I_2(g) \rightleftharpoons 2HI(g)$ __________

5. What is the value of ΔG^o if at 25°C a reaction has an equilibrium constant equal to 1.0? __________

6. What is the value of ΔG^o if at 25°C a reaction has an equilibrium constant of 1.5×10^{-12}? __________

New Terms

thermodynamic equilibrium constant

12.5 THE RELATIONSHIP BETWEEN K_p AND K_c

Objectives

To see how we can convert from K_p to K_c, and vice versa.

Review

Remember the relationship,

$$K_p = K_c(RT)^{\Delta n_g}$$

where Δn_g is the change in the number of moles of gas on going from reactants to products in the balanced equation. Review the sample calculation in Example 12.4 in the text. Notice that in this calculation R = 0.0821 l atm/mole K. This is to make units cancel properly.

Self-Test

7. The reaction, $N_2(g) + 3H_2(g) \rightleftharpoons 2NH_3(g)$, has $K_p = 7.2 \times 10^5$ at 25^oC. What is the value of K_c? ____________

8. The reaction, $H_2(g) + I_2(g) \rightleftharpoons 2HI(g)$, has $K_p = 0.35$ at 25^oC. What is the value if K_c? ____________

New Terms

none

12.6 HETEROGENEOUS EQUILIBRIA

Objectives

To see how the mass action expression can be simplified in cases of equilibrium between two or more pure phases.

Review

Remember that the concentrations of pure liquid or solid phases are not included in the mass action expression. This is because they are constant and are included in the equilibrium constant.

Self-Test

9.   Write the equilibrium constant expression for $\underline{K}_c$ for the following:

(a) $2H_2(g) + O_2(g) \rightleftharpoons 2H_2O(l)$ _____

(b) $CO_2(g) + Li_2CO_3(s) + H_2O(g) \rightleftharpoons 2LiHCO_3(s)$

_____

(c) $Cl_2(g) + 2KBr(s) \rightleftharpoons 2KCl(s) + Br_2(g)$

_____

New Terms

heterogeneous equilibrium

12.7 LE CHATELIER'S PRINCIPLE AND CHEMICAL EQUILIBRIUM

Objectives

To learn how to apply Le Chatelier's principle to changes in concentration, pressure and temperature in chemical systems.

Review

(1) When a reactant or product is added to a system at equilibrium, the position of equilibrium shifts toward the opposite side of the equation.

(2) Decreasing the concentration of a reactant or product causes the position of equilibrium to shift in the direction of the substance removed.

(3) Increasing the pressure by decreasing the volume shifts the position of equilibrium in the direction of the fewest number of moles of gas.

(4) An increase in temperature causes the position of equilibrium to shift in the direction of the endothermic reaction.

A very important point to remember is that the only thing that changes $\underline{K}$ for a reaction is a change in temperature!

Two final observations are made in this section. One is that adding an inert (unreactive) gas to a system, without changing the volume, has no effect on the position of equilibrium. The second is that a catalyst has no effect on the position of equilibrium. It only increases the speed with which the system reaches equilibrium.

Self-Test

10. Use the letters, I = increase, D = decrease, N = no change, to indicate what effect each of the following changes will havc upon the concentration of $SO_2(g)$ in the system,

$$2SO_2(g) + O_2(g) \rightleftharpoons 2SO_3(g) \qquad \Delta H = -46.2 \text{ kcal}$$

(a) adding $O_2(g)$ ________

(b) adding $SO_3(g)$ ________

(c) removing $SO_3(g)$ ________

(d) increasing the temperature ________

(e) increasing the volume of the container ________

(f) adding helium ________

(g) adding a catalyst ________

11. Which, if any, of the changes described in Question 13 will alter the equilibrium constant? ________________

New Terms

none

12.8 EQUILIBRIUM CALCULATIONS

Objectives

To learn how to use the equilibrium constant expression in numerical calculations.

Review

There are basically two kinds of calculations that you have to learn to do. One is to calculate K from either equilibrium concentrations or information from which you can deduce equilibrium concentrations. The other is to calculate information about equilibrium concentrations, having at your disposal the value of K. There are six sample calculations in the text illustrating these types of problems (Examples 12.6 to 12.11). These are worked out in great detail. Two additional sample calculations follow. A very important thing to notice in all of these is that the concentrations (or algebraic quantities representing concentrations) that are substituted into the mass action expression are always equilibrium concentrations.

The first example deals with the calculation of $\underline{K}$ from a set of equilibrium concentrations.

Example 12.2

At a certain temperature, 0.15 mole CO(g) and 0.15 mole $H_2O(g)$ are introduced into a 2.0-liter container. At equilibrium the CO(g) concentration was measured to be 0.042 mole/liter. What is the value of $\underline{K}_c$ for the reaction,

$$CO(g) + H_2O(g) \rightleftharpoons H_2(g) + CO_2(g)$$

Solution

First let's write the equilibrium constant expression.

$$\underline{K}_c = \frac{[H_2][CO_2]}{[CO][H_2O]}$$

We need equilibrium concentrations of each gas. First let's calculate the initial concentrations.

$$[CO] = 0.15 \text{ mole}/2.0 \text{ liter} = 0.075 \text{ M}$$
$$[H_2O] = 0.15 \text{ mole}/2.0 \text{ liter} = 0.075 \text{ M}$$

How about the products? Their initial concentrations were zero. At equilibrium they are present because some CO and H_2O reacted to produce them. How much reacted?

We are given the equilibrium concentration of CO, 0.042 M. This is less than we started with. The difference between what is present at equilibrium and what we started with is the amount that reacted.

$$\text{Amount of CO reacted} = 0.075 \text{ mole/liter} - 0.042 \text{ mole/liter}$$
$$= 0.033 \text{ mole/liter}$$

How much H_2O reacted with the CO? From the equation we can see that the answer must also be 0.033 mole/liter, and the amount of H_2O remaining at equilibrium must be

$$[H_2O]_{\text{equilibrium}} = 0.075 \text{ M} - 0.033 \text{ M} = 0.042 \text{ M}$$

For the reactants at equilibrium,

$$[H_2O] = 0.042 \text{ M}$$
$$[CO] = 0.042 \text{ M}$$

How about the products? When 0.033 mole/liter of CO reacts, it must produce 0.033 mole/liter of both H_2 and CO_2. This we can see from the coefficients in the balanced equation. At equilibrium, then,

$$[H_2] = 0.033 \text{ M}$$
$$[CO_2] = 0.033 \text{ M}$$

Substituting these equilibrium concentrations into the mass action expression, we can calculate $\underline{K}$.

$$\underline{K}_c = \frac{(0.033)(0.033)}{(0.042)(0.042)}$$

$$\underline{K}_c = 0.62$$

The second example asks you to calculate equilibrium concentrations using the known value of $\underline{K}$. This kind of question involves some very simple algebra. Don't panic over it. If you take your time and think it through slowly, you should be able to learn how to approach this kind of problem. Don't try to memorize how to solve specific problems. If you do, it only takes a small change in the problem to trip you up.

Example 12.3

In the last example we found that at a particular temperature $\underline{K}_c$ = 0.62 for the reaction,

$$CO(g) + H_2O(g) \rightleftharpoons H_2(g) + CO_2(g)$$

The equilibrium concentrations were: $[CO] = [H_2O] = 0.042$ M and $[H_2] = [CO_2] = 0.033$ M. Suppose an additional 0.010 mole/liter of CO(g) and 0.010 mole/liter of H_2O(g) are introduced into the container. What will the new equilibrium concentrations become?

Solution

As soon as the additional CO and H_2O are added, the equilibrium is destroyed and we can treat these as initial concentrations.

initial concentrations

$$[CO] = 0.042 + 0.010 = 0.052 \text{ M}$$
$$[H_2O] = 0.042 + 0.010 = 0.052 \text{ M}$$
$$[H_2] = 0.033 \text{ M}$$
$$[CO_2] = 0.033 \text{ M}$$

We now take into account that some reaction will take place. Let's let x equal the number of moles/liter of CO that will react. The CO concentration at equilibrium will therefore have been diminished by x. The H_2O concentration will also have been decreased by x mole/liter (from the balanced equation). Similarly, the H_2 and CO_2 will each increase by x.

<u>equilibrium concentrations</u>

$$[CO] = 0.052 - x$$
$$[H_2O] = 0.052 - x$$
$$[H_2] = 0.033 + x$$
$$[CO_2] = 0.033 + x$$

The reasoning that we have just gone through is the key to solving the problem. Make sure you understand it. Now, substituting these equilibrium quantities into the mass action expression (see the last example),

$$0.62 = \frac{(0.033 + x)(0.033 + x)}{(0.052 - x)(0.052 - x)} = \frac{(0.033 + x)^2}{(0.052 - x)^2}$$

The simplest way to solve this problem is to take the square root of both sides of the equation; $\sqrt{0.62} = 0.79$

$$0.79 = \frac{0.033 + x}{0.052 - x}$$

Now multiply both sides by (0.052 - x).

$$0.79(0.052 - x) = 0.033 + x$$
$$0.041 - 0.79x = 0.033 + x$$
$$0.041 - 0.033 = x + 0.79x$$
$$0.008 = 1.79x$$
$$0.004 = x$$

The equilibrium concentrations become

$$[CO] = [H_2O] = 0.052 - 0.004 = 0.048\ M$$
$$[H_2] = [CO_2] = 0.033 + 0.004 = 0.037\ M$$

<u>Self-Test</u>

12. At approximately 1700°C the equilibrium concentrations of the reactants and product in the equation,

$$N_2(g) + O_2(g) \rightleftharpoons 2NO(g)$$

are $[N_2] = 1.0 \times 10^{-4}$ M, $[O_2] = 2.5 \times 10^{-5}$ M
$[NO] = 7.1 \times 10^{-7}$ M
What is $\underline{K}_c$ for this reaction? ________________

13. In a furnace operating at 1700°C the concentrations of N_2 and O_2 are 4.0×10^{-4} M and 1.0×10^{-5} M, respectively. At this temperature $\underline{K}_c = 2.0 \times 10^{-4}$ (the answer to the last question!). What concentration of NO(g) will be present in the gases escaping from the furnace?

(Assume the N_2 and O_2 concentrations are equilibrium concentrations)

_____

14. At a certain temperature $K_c = 0.5$ for the reaction,

$$H_2(g) + I_2(g) \rightleftharpoons 2HI(g)$$

If 0.40 mole of H_2 and 0.40 mole of I_2 are placed into a 1.0-liter container at this temperature, what will be the equilibrium concentration of each gas?

_____

New Terms

ANSWERS TO SELF-TEST QUESTIONS

1. (a)
$$\underline{K}_c = \frac{[PCl_3][Cl_2]}{[PCl_5]} \qquad \underline{K}_p = \frac{p_{PCl_3}\, p_{Cl_2}}{p_{PCl_5}}$$

(b)
$$\underline{K}_c = \frac{[H_2SO_4][HBr]^2}{[Br_2][SO_2][H_2O]^2} \qquad \underline{K}_p = \frac{p_{H_2SO_4}\, p_{HBr}^2}{p_{Br_2}\, p_{SO_2}\, p_{H_2O}^2}$$

(c)
$$\underline{K}_c = \frac{[XeO_3]^2 [Xe]^4 [O_2][HF]^{24}}{[XeF_4]^6 [H_2O]^{12}} \qquad \underline{K}_p = \frac{p_{XeO_3}^2\, p_{Xe}^4\, p_{O_2}\, p_{HF}^{24}}{p_{XeF_4}^6\, p_{H_2O}^{12}}$$

(d)
$$\underline{K}_c = \frac{[NO_2F]^2}{[NO_2]^2[F_2]} \qquad \underline{K}_p = \frac{p_{NO_2F}^2}{p_{NO_2}^2\, p_{F_2}}$$

2. K = 217 when values are substituted in the expression,
$$K = \frac{[NO_2]^2}{[N_2O_4]}$$

3. d < b < a < c   4. K = 0.35   5. zero, since log 1.0 = 0
6. +16.1 kcal (+67.5 kJ)   7. $\Delta n_g = -2$, $\underline{K}_c = 4.3 \times 10^8$ 8. $\Delta n_g = 0$,
$\underline{K}_c = \underline{K}_p = 0.35$ 9. (a) $\underline{K}_c = \dfrac{1}{[H_2]^2[O_2]}$ (b) $\underline{K}_c = \dfrac{1}{[CO_2][H_2O]}$

(c) $\underline{K}_c = \dfrac{[Br_2]}{[Cl_2]}$

10. (a) D (b) I (c) D (d) I (e) I (f) N (g) N
11. only (d) will change K
12. $K_c = 2.0 \times 10^{-4}$
13. 8.9×10^{-7} M
14. $[H_2] = [I_2] = 0.30$ M; $[HI] = 0.20$ M

13 ACIDS AND BASES

The acid-base concept is one of the most useful ways of correlating a large amount of what otherwise might appear to be unrelated chemical information. In this chapter the acid-base definition presented in Chapter 5 is expanded to cover acid-base reactions in solvents other than water, and even to reactions in the absence of any solvent whatsoever. Before reading the first section, review the three properties that acids and bases have in common, which are given in the introduction to the chapter in the text.

13.1 THE ARRHENIUS DEFINITION OF ACIDS AND BASES

Objectives

To review the definition of acids and bases in the solvent, water. You should learn what characterizes an acid and a base in water, and you should learn the neutralization reaction that occurs in aqueous solutions.

Review

Remember, in water an acid produces H_3O^+ and a base produces OH^-. Many acids contain hydrogen; for example, HCl, H_2SO_4, HNO_3. Other substances react with water to produce acids; these are called acid anhydrides. Bases often contain OH^- (e.g., NaOH, $Ba(OH)_2$). Some react with water to generate OH^-. An example is NH_3. In general,

nonmetal oxides are acid anhydrides
metal oxides are basic anhydrides

Self-Test

1. Write the chemical equation for the reaction of the following with water:

   (a) HBr ______________________

   (b) H_2SO_4 ______________________

   ____

   (c) N_2H_4 ______________________

   (d) O^{2-} ______________________

2. Identify the following as acid (A) or basic (B) anhydrides.

   (a) CaO ________ (d) Li_2O ________

   (b) SO_2 ________ (e) N_2O_3 ________

   (c) CO_2 ________ (f) P_4O_6 ________

3. Write chemical equations showing the reactions of the following with water.

   (a) SO_3 ______________________

   (b) P_4O_{10} ______________________

   (c) BaO ______________________

   (d) N_2O_3 ______________________

4. What is the net neutralization reaction between an acid and a base in aqueous solution?

   ____

5. What three general properties do acids and bases exhibit?

   (a) ______________________

   (b) ______________________

   (c) ______________________

New Terms

acid anhydride
basic anhydride

13.2 BRØNSTED-LOWRY DEFINITION OF ACIDS AND BASES

Objectives

To define acids and bases in terms of the transfer of a proton. You should learn to identify acid-base conjugate pairs in an acid-base reaction. You should learn what an amphiprotic (amphoteric) substance is.

Review

Learn the Brønsted-Lowry definitions of acid and base:

acid - proton (H^+) donor
base - proton acceptor

Acid-base reactions can be looked upon as reversible.

$$\text{Acid (X)} + \text{Base (Y)} \rightleftharpoons \text{Base (X)} + \text{Acid (Y)}$$

$$HX + Y \rightleftharpoons X^- + HY^+$$

There are two conjugate acid-base pairs in this reaction.

Acid (X) - Base (X)		(HX, X^-)
Base (Y) - Acid (Y)		(Y, HY^+)

Let's look at a concrete example,

$$HNO_3 + H_2O \rightleftharpoons H_3O^+ + NO_3^-$$

acid base acid base

(H_2O base and H_3O^+ acid: conjugate pair)

(HNO_3 acid and NO_3^- base: conjugate pair)

Notice that the only difference between the members of a conjugate pair is a single proton. All other atoms are identical. Also, note that the acid has one more hydrogen than the base.

Example 13.1

From the list below, choose those two which form a conjugate acid-base pair. Which one is the acid?

$$H_2SO_4,\ OH^-,\ HSO_3^-,\ SO_3^{2-},\ SO_4^{2-}$$

Solution

The only two species that differ from each other by only one hydrogen are

$$HSO_3^- \text{ and } SO_3^{2-}$$

The one with the most hydrogen (HSO_3^-) is the acid.

An amphiprotic (amphoteric)substance can act as either an acid or a base. The most common example is water.

acid	base	
H_3O^+	H_2O	OH^-
	acid	base

Review the autoionization reactions on Page 377 of the text. Also, note that the Brønsted-Lowry definition permits acid-base reactions in the absence of a solvent. The reaction between HCl(g) and NH_3(g) is an example.

Self-Test

6. In each of the following indicate whether the underlined substance is behaving as an acid (A) or a base (B).

   (a) $H_2O + \underline{HC_2H_3O_2} \rightleftharpoons H_3O^+ + C_2H_3O_2^-$ ________

   (b) $\underline{CN^-} + HCl \rightleftharpoons HCN + Cl^-$ ________

   (c) $\underline{NH_3} + O^{2-} \rightleftharpoons NH_2^- + OH^-$ ________

7. For each part in Question 6, identify both acid-base conjugate pairs. Underline the acid in each of them.

   (a) ______________________, ______________________

   (b) ______________________, ______________________

   (c) ______________________, ______________________

8. Which of the following could not serve as amphiprotic substances?

   (a) NH_3 (b) CN^- (c) O^{2-} (d) HSO_4^- (e) SO_4^{2-} ________________

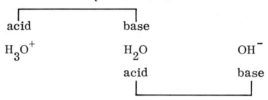

New Terms

conjugate acid
conjugate base
conjugate acid-base pair
amphiprotic
amphoteric

13.3 STRENGTHS OF ACIDS AND BASES

Objectives

To compare the relative strengths of acids and bases.

Review

The position of equilibrium in the reaction,

$$HA + B^- \rightleftharpoons HB + A^-$$

allows us to compare the relative strengths of the acids and bases. A strong acid tends to give up its proton more readily than a weak acid. If HA is stronger than HB, the position of equilibrium lies to the right. Similarly, a strong base is able to capture protons more readily than a weak base. If B^- is a stronger base than A^-, the B^- captures more protons than the A^- and the position of equilibrium again lies to the right. In general the position of equilibrium lies in the direction of the weaker acid and base.

$$\text{strong acid} + \text{strong base} \rightleftharpoons \text{weak acid} + \text{weak base}$$

The leveling effect occurs when comparing the strengths of strong acids in a basic solvent, or strong bases in an acidic solvent. Any solvent that makes it possible to distinguish between the strengths of acids or bases is called a differentiating solvent.

Self-Test

9. The following are equations for the reaction of some Brønsted acids with water. The equilibrium constants are also given. If necessary, review the relationship between position of equilibrium and the magnitude of K as given in Section 12.2 (see Page 180 in the Study Guide and Page 354 in the text). Arrange the acids in order of increasing acid strength.
   (a) $HNO_2 + H_2O \rightleftharpoons H_3O^+ + NO_2^-$ $K = 4.5 \times 10^{-4}$
   (b) $HF + H_2O \rightleftharpoons H_3O^+ + F^-$ $K = 6.5 \times 10^{-4}$
   (c) $HCN + H_2O \rightleftharpoons H_3O^+ + CN^-$ $K = 4.9 \times 10^{-10}$

(d) $H_3PO_4 + H_2O \rightleftharpoons H_3O^+ + H_2PO_4^-$ $K = 7.5 \times 10^{-3}$

(e) $HC_3H_5O_2 + H_2O \rightleftharpoons H_3O^+ + C_3H_5O_2^-$ $K = 1.4 \times 10^{-5}$

____

10. The ions, Br^- and Cl^-, are very weak Brønsted bases. What property would a solvent have to have to be considered a differentiating solvent for these ions? ____

New Terms

leveling effect
leveling solvent
differentiating solvent

13.4 FACTORS INFLUENCING THE STRENGTHS OF ACIDS

Objectives

To understand why some acids are stronger than others. You should be able to predict the relative acid strengths among oxoacids (acids containing hydrogen, a nonmetal, and oxygen) and among hydro acids (acids containing only hydrogen and a nonmetal).

Review

Below are summarized the trends in acidity that we observe among different kinds of acids. You should be aware of these, and the Self-Test provides a way to test your ability to apply the rules. In addition, you should try to understand why these trends occur. If you're not sure about the reasons, reread the explanations provided in the text.

Oxoacids, H_nXO_m

(1) Strength increases as the number of lone oxygen atoms increases.

(2) For acids of the same general formula, strength decreases as the central atom gets larger (as you descend a group).

(3) In general, attaching an electronegative atom (or group of atoms) to an atom bonded to an O-H group increases the acidity of the O-H group.

(4) An increase in the oxidation number of an atom that is bonded to an O-H group increases the acidity of the O-H group.

Hydro acids (binary acids), H_nX

(1) Strength increases from left to right within a period.

(2) Strength increases from top to bottom within a group.

Self-Test

11. In each pair below, choose the strongest acid.

(a) H_2SeO_3, H_2SeO_4 ______________

(b) $HBrO_3$, HBrO ______________

(c) HNO_3, HNO_2 ______________

(d) H_3PO_4, H_2SO_4 ______________

(e) H_3PO_4, H_3AsO_4 ______________

(f) H_3N, H_2O ______________

(g) PH_3, NH_3 ______________

(h) H_2Se, HBr ______________

12. What are the meanings of the following terms?

(a) solvolysis ______________________________

(b) hydrolysis ______________________________

New Terms

solvolysis
hydrolysis

13.5 LEWIS ACIDS AND BASES

Objectives

To provide a still more general definition of an acid and a base. You should learn how the Lewis definition can be used to explain acid-base reactions.

Review

Under the Lewis definition we have:

base – electron pair donor
acid – electron pair acceptor

A Lewis acid and base react with each other by the formation of a coordinate covalent bond. Lewis bases tend to be species that have lone pairs of electrons and completed octets. Lewis acids tend to be substances that can be considered electron deficient, or which can make themselves electron deficient by rearrangement of their electrons. The latter is illustrated by SO_3 when it reacts with oxide ion.

Lewis acid–base reactions are often viewed as displacement reactions.

(1) nucleophilic displacement – one base displaces another.

$$[:\ddot{\underset{..}{S}}:]^{2-} \ + \ \overset{\text{acid}}{\underset{H}{\overset{H}{\diagup \ :\ddot{O}:}}} \longrightarrow [:\ddot{\underset{..}{S}} - H]^{-} \ + \ [:\ddot{\underset{..}{O}} - H]^{-}$$

incoming   outgoing
base   base
(nucleophile)

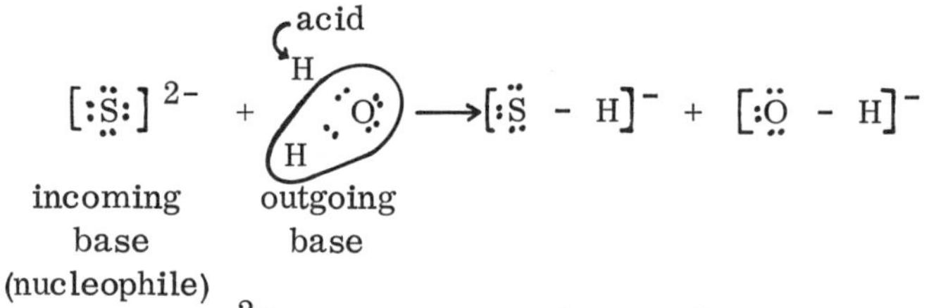

$$S^{2-} + H_2O \longrightarrow HS^- + OH^-$$

(2) electrophilic displacement – one acid displaces another.

$$Cl_5Sb \ + \ \overset{:\ddot{Cl}:}{\underset{:O:}{\overset{..}{N:}}} \longrightarrow [Cl_5Sb : \ddot{Cl}:]^- \ + \ [:N \!\!=\!\! O:]^+$$

incoming   outgoing
acid   acid
(electrophile)

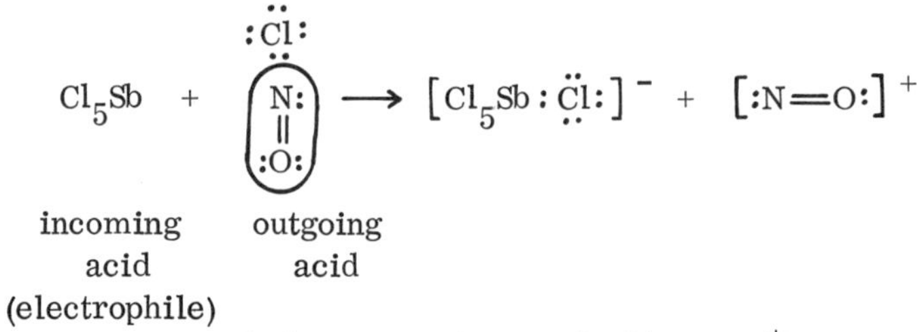

$$SbCl_5 \ + \ NOCl \longrightarrow SbCl_6^- \ + \ NO^+$$

Study the rules given on Page 387 that explain how you can identify nucleophilic and electrophilic displacement reactions.

Self–Test

13. Use electron–dot formulas to show how the reaction of O^{2-} with SO_2 to produce SO_3 can be considered the reaction between a Lewis acid and base.

14. Identify the following reactions as nucleophilic (N) or electrophilic (E) displacements.

(a) $PCl_5 + PCl_5 \longrightarrow PCl_4^+ + PCl_6^-$ ________

(b) $(CH_3)_3CBr + OH^- \longrightarrow (CH_3)_3COH + Br^-$ ________

(c) $B(OH)_3 + H_2O \longrightarrow B(OH)_4^- + H^+$ ________

(d) $Cu(H_2O)_4^{2+} + Cl^- \longrightarrow Cu(H_2O)_3Cl^+ + H_2O$ ________

New Terms

Lewis acid
Lewis base
nucleophile
nucleophilic displacement
electrophile
electrophilic displacement

13.6 THE SOLVENT SYSTEM APPROACH TO ACIDS AND BASES

Objectives

To define acids and bases in terms of the ions produced by the autoionization of the solvent.

Review

According to the solvent system approach we have:

acid - substances that increase the concentration of the solvent cation.
base - substances that increase the concentration of the solvent anion.

The usefulness of this approach is that it permits reasoning by analogy from one solvent to another. Compare, for example, neutralization reactions in water and ammonia.

Self-Test

15. Bromine trifluoride, BF_3, undergoes autoionization,

$$BF_3 + BF_3 \rightleftharpoons BF_2^+ + BF_4^-$$

There are also the reactions,

$$SbF_5 + BF_3 \longrightarrow SbF_6^- + BF_2^+$$

$$KF + BF_3 \longrightarrow K^+ + BF_4^-$$

How would you classify (acid or base) the following?

(a) SbF_5 ____________ (b) KF ____________

16. In some situations, liquid sulfur dioxide behaves as if it undergoes autoionization,

$$SO_2 + SO_2 \rightleftharpoons SO^{2+} + SO_3^{2-}$$

How would you expect the following to behave in liquid SO_2?

(a) Na_2SO_3 __________ (b) $SOCl_2$ __________

(c) What reaction, if any, would occur between Na_2SO_3 and $SOCl_2$ in liquid SO_2? ____________________

New Terms

solvent system

13.7 SUMMARY

Objectives

To bring into perspective the features of the different acid-base definitions discussed in this chapter.

New Terms

none

ANSWERS TO SELF-TEST QUESTIONS

1. (a) $HBr + H_2O \longrightarrow H_3O^+ + Br^-$

   (b) $H_2SO_4 + H_2O \longrightarrow H_3O^+ + HSO_4^-$

   $HSO_4^- + H_2O \longrightarrow H_3O^+ + SO_4^{2-}$

   (c) $N_2H_4 + H_2O \longrightarrow N_2H_5^+ + OH^-$

   (d) $O^{2-} + H_2O \longrightarrow 2OH^-$

2. (a) B (b) A (c) A (d) B (e) A (f) A

3. (a) $SO_3 + H_2O \longrightarrow H_2SO_4$

   (b) $P_4O_{10} + 6H_2O \longrightarrow 4H_3PO_4$

   (c) $BaO + H_2O \longrightarrow Ba(OH)_2$

   (d) $N_2O_3 + H_2O \longrightarrow 2HNO_2$

4. $H_3O^+ + OH^- \longrightarrow 2H_2O$, or simply, $H^+ + OH^- \longrightarrow H_2O$

5. (a) neutralization (b) reaction with indicators (c) catalysis

6. (a) acid (b) base (c) acid

7. (a) H_2O, $\underline{H_3O^+}$; $\underline{HC_2H_3O_2}$, $C_2H_3O_2^-$

(b) CN^-, $\underline{HCN}$; $\underline{HCl}$, Cl^-

(c) $\underline{NH_3}$, NH_2^-; O^{2-}, $\underline{OH^-}$

8. b, c and e. They can't be Brønsted acids because they have no hydrogen.

9. $HCN < HC_3H_5O_2 < HNO_2 < HF < H_3PO_4$

10. It would have to be a very acidic solvent so that some protonation of each would occur. The weaker of the two would tend to be protonated the least. 11. (a) H_2SeO_4 (b) $HBrO_3$ (c) HNO_3 (d) H_2SO_4 (e) H_3PO_4 (f) H_2O (g) PH_3 (h) HBr 12. (a) solvolysis - reaction of a substance with the solvent (b) hydrolysis - reaction of a substance with water.

13.

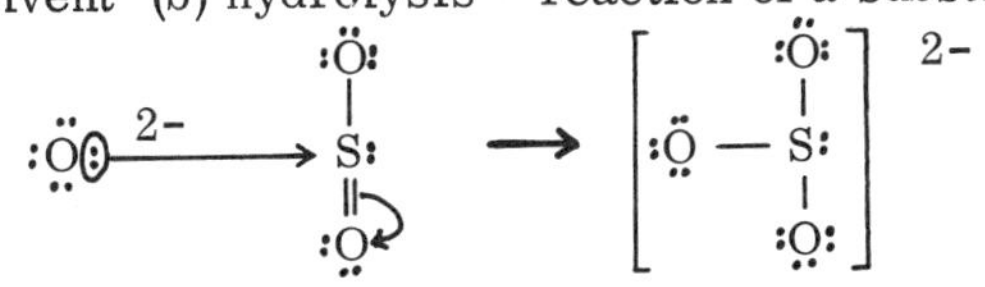

sulfite ion

14. (a) E, Cl^- is the base; PCl_4^+ is the acid that is displaced. (b) N, $(CH_3)_3C^+$ is the acid; Br^- is the base displaced by OH^-. (c) E, OH^- is the base in H_2O; H^+ is the acid that is displaced. The substance $B(OH)_3$ is boric acid. (d) N, $Cu(H_2O)_3^{2+}$ is the acid; H_2O is the base displaced by Cl^-.

15. (a) acid - it increases BF_2^+ (b) base - it increases BF_4^-

16. (a) base (b) acid, $SOCl_2 \longrightarrow SO^{2+} + 2Cl^-$ (c) These react together, $Na_2SO_3 + SOCl_2 \longrightarrow 2SO_2 + 2NaCl$ This is a neutralization reaction in liquid SO_2.

14 ACID-BASE EQUILIBRIA IN AQUEOUS SOLUTION

This chapter deals quantitatively with the equilibria involving the autoionization of water and the dissocaition of weak acids and bases. These are important in any aqueous system, particularly biological ones where many important molecules behave as weak acids or bases. Many of the numerical problems in this chapter and the next require the application of some simple algebra. Don't panic ! Follow the procedures shown in the worked-out examples which explain how to approach these problems. The important thing is to try not to rush - don't skip steps in the reasoning. If you proceed slowly, you should be able to master the material in Chapters 14 and 15.

14.1 IONIZATION OF WATER, pH

Objectives

To establish the quantitative criteria for equilibrium in the autoionization of water and to devise a system for expressing small concentrations of H_3O^+.

Review

In this section you saw that the equilibrium condition for the ionization of water reduces to

$$K_w = [H^+][OH^-] = 1.0 \times 10^{-14}$$

In any solution in which water is the solvent this condition holds. You are expected to know the value, $K_w = 1.0 \times 10^{-14}$.

In pure water, $[H^+] = [OH^-] = 1.0 \times 10^{-7}$ M. When H^+ or OH^- are present from another source (e.g., an acid or a base in the solution), $[H^+] \neq [OH^-]$. The product of concentrations, however, must equal K_w.

In general, for any quantity X,

$$pX = -\log X$$

For the hydrogen in concentration,

$$pH = -\log [H^+]$$

Similarly, for hydroxide ion,

$$pOH = -\log [OH^-]$$

Remember that the sum of pH and pOH equals pK_w.

$$pH + pOH = pK_w = 14.0$$

You should be able to calculate pH given $[H^+]$, and also $[H^+]$ given pH. Review Examples 14.2, 14.3 and 14.4 in the text. To solve these problems it is necessary to refer to the table of logarithms (Appendix D) or use an electronic calculator having a LOG function. Actually, the availability of electronic calculators has greatly simplified pH calculations. If you have a calculator with a LOG key, the pH can be found by entering the $[H^+]$ and depressing the LOG key. The pH, preceded by a minus sign, will appear on the display. To obtain the $[H^+]$ from pH you can use the Y^X key, because $[H^+] = 10^{-pH}$. Enter 10 as the value of Y, depress the Y^X key, then enter the negative of the pH (i.e., -pH) as the value of X. Finally, depress the "=" key and the $[H^+]$ will appear on the display. The actual sequence of operations, of course, may differ somewhat for different brands of calculators.

Self-Test

1. Calculate the $[OH^-]$ in the following solutions:

   (a) $[H^+] = 1.0 \times 10^{-8}$ M ____________

   (b) $[H^+] = 4.2 \times 10^{-12}$ M ____________

   (c) $[H^+] = 8.4 \times 10^{-2}$ M ____________

2. Calculate the $[H^+]$ in the following solutions:

   (a) $[OH^-] = 1.0 \times 10^{-4}$ M ____________

   (b) $[OH^-] = 2.8 \times 10^{-9}$ M ____________

   (c) $[OH^-] = 6.7 \times 10^{-12}$ M ____________

3. Calculate the pH and pOH of solutions with the following concentrations:

   (a) $[H^+] = 2.0 \times 10^{-6}$ M ________________

   (b) $[H^+] = 8.5 \times 10^{-9}$ M ________________

   (c) $[OH^-] = 3.4 \times 10^{-6}$ M ________________

4. Calculate the $[H^+]$ in the following solutions:

   (a) pH = 8.50 ________________

   (b) pH = 13.34 ________________

   (c) pOH = 13.34 ________________

New Terms

ion product constant
ionization constant
dissociation constant
pH

14.2 DISSOCIATION OF WEAK ELECTROLYTES

Objectives

To deal quantitatively with equilibria involving the dissociation of weak electrolytes, in particular weak acids and bases. You should be able to calculate K, given equilibrium concentrations. You should be able to calculate equilibrium concentrations from K and the concentration of the weak electrolyte.

Review

For a weak acid that undergoes the ionization,

$$HA \rightleftharpoons H^+ + A^-$$

we have

$$K_a = \frac{[H^+][A^-]}{[HA]}$$

For a weak base, B, that reacts with the solvent,

$$B + H_2O \rightleftharpoons HB^+ + OH^-$$

we have

$$K_b = \frac{[HB^+][OH^-]}{[B]}$$

Examples 14.5 and 14.6 in the text show you how to calculate the equilibrium constant if you have information that allows you to compute equilibrium concentrations.

To calculate equilibrium concentrations from K you must know the amount of weak acid or base placed in the solution. The following example is typical.

Example 14.1

What are the concentrations of H^+, OCl^- and HOCl in a solution labeled 0.10 M HOCl? For HOCl, $K_a = 3.1 \times 10^{-8}$.

Solution

First, write the equation for the equilibrium.

$$HOCl \rightleftharpoons H^+ + OCl^-$$

The next step for problems of this type is to set up our table of concentrations as shown in the example problems in the text.

	initial concentration	change	equilibrium concentration
H^+	0.0	+x	x
OCl^-	0.0	+x	x
HOCl	0.10	-x	0.10 - x
	This column contains the concentrations of solutes placed into the solution.	This column represents the change that occurs because of reaction.	This column gives the equilibrium concentrations.

The entries in the table are obtained by the following reasoning. If no dissociation occurred, we would have [HOCl] = 0.10 M. Some does dissociate, however, and we want to calculate how much. The approach, then, is to let x = no. of moles per liter of HOCl that dissociate when equilibrium is reached. The concentrations at equilibrium, then, would be

$[HOCl] = 0.10 - x$ (note that the concentration of HOCl is diminished by the quantity that is lost upon dissociation)

No H^+ or OCl^- were placed in the solution; they are there at equilibrium because of the dissociation of the HOCl.

$[OCl^-] = x$ (note that $[OCl^-] = [H^+]$ because they are formed in a 1:1 ratio. The amount produced is equal to the amount of HOCl that is dissociated)

$[H^+] = x$

The equilibrium constant expression is

$$K_a = \frac{[H^+][OCl^-]}{[HOCl]} = 3.1 \times 10^{-8}$$

Substituting our expressions for the equilibrium concentrations from the last column of the table gives

$$\frac{(x)(x)}{(0.10 - x)} = 3.1 \times 10^{-8}$$

Expanding this out will give a quadratic equation. The problem can be simplified however. From the magnitude of K_a we know that very little HOCl will dissociate. What we do is assume that the amount that dissociates is negligible; that is,

$$0.10 - x \approx 0.10$$

This simplification gives

$$\frac{(x)(x)}{(0.10)} = 3.1 \times 10^{-8}$$

$$x^2 = (0.10)(3.1 \times 10^{-8}) = 3.1 \times 10^{-9}$$

$$x^2 = 31 \times 10^{-10}$$

$$x = \sqrt{31} \times 10^{-5}$$

$$x = 5.6 \times 10^{-5}$$

(Note that to take the square root without an electronic calculator, you must have the exponent on 10 divisible by 2.)

Finally, we have the equilibrium concentrations,

$$[HOCl] = 0.10 - 5.6 \times 10^{-5} = 0.10 \text{ M}$$

$$[H^+] = [OCl^-] = 5.6 \times 10^{-5} \text{ M}$$

Observe that x is indeed negligible compared to 0.10 when the difference is rounded to the proper number of significant figures. This justifies our assumption.

Sometimes you will have to deal with problems involving solutions containing a weak acid or base plus a salt of that acid or base. An example is $HC_2H_3O_2$ and $NaC_2H_3O_2$. In working problems of this type you should remember the following:

(1) salts are completely dissociated. The concentration of the ion produced by the salt is entered in the "initial concentration" column.

(2) in the problems that you will encounter only one of the ions produced by the salt is important. The other is a spectator ion.

Review Example 14.8 in the text.

Self-Test

5. A 0.20 M solution of propionic acid, $HC_3H_5O_2$, has a hydrogen ion concentration of 1.7×10^{-3} M. What is the value of K_a for propionic acid? ____________

6. A 1.0 molar solution of the weak base methyl amine, CH_3NH_2, has a pH of 12.32. What is the value of K_b? ____________

7. Aniline, $C_6H_5NH_2$, has an ionization constant, $K_b = 3.8 \times 10^{-10}$. Aniline reacts with water according to the equation, $C_6H_5NH_2 + H_2O \rightleftharpoons C_6H_5NH_3^+ + OH^-$. What are the OH^- and H^+ concentrations in a 0.01 M solution of $C_6H_5NH_2$? ____________

____

8. Calculate the percent dissociation in a 0.25 M solution of HCN. $K_a = 4.9 \times 10^{-10}$ ____________

9. What is the value of pK_a for HCN? K_a is given in Question 8. ____________

10. What is the pH of a 0.20 M solution of lactic acid, $HC_3H_5O_3$, that also contains 0.30 M $NaC_3H_5O_3$? For lactic acid, $K_a = 1.38 \times 10^{-4}$. ____________

New Terms

acid dissociation constant

14.3 DISSOCIATION OF POLYPROTIC ACIDS

Objectives

To consider equilibria for acids that dissociate in two steps. You should learn how to calculate the hydrogen ion concentration as well as the concentration of other ions produced in the equilibria.

Review

For a diprotic acid there are two equilibria and two equilibrium constants. For H_2S

$$H_2S \rightleftharpoons H^+ + HS^- \qquad K_{a1} = \frac{[H^+][HS^-]}{[H_2S]}$$

$$HS^- \rightleftharpoons H^+ + S^{2-} \qquad K_{a2} = \frac{[H^+][S^{2-}]}{[HS^-]}$$

Remember, to calculate the hydrogen ion concentration in a solution of a polyprotic acid you always use K_{a1}. This is because the second and succeeding ionization steps almost always produce much less H^+ than the first (i.e., $K_{a1} \gg K_{a2} \gg K_{a3} \ldots$). You should also use K_{a1} to calculate the concentration of the anion produced in the first step.

For diprotic acids you should use K_{a2} to calculate the anion produced in the second dissociation (e.g., S^{2-} in the second ionization of H_2S). If the diprotic acid is the only solute in the solution, the concentration of the anion, A^{2-} (e.g., S^{2-}) is equal to K_{a2}.

In a solution of a diprotic acid, H_2A, a combined expression

$$K_{a1}\,K_{a2} = \frac{[H^+]^2[A^{2-}]}{[H_2A]}$$

can be used provided that the values of any two of the three concentrations that appear in the expression are known. These are the only conditions under which this combined expression can be applied.

Self-Test

11. Calculate the concentrations of the ions, H^+, HA^- and A^{2-}, in a 0.10 M solution of the weak acid, H_2A. $K_{a1} = 2.0 \times 10^{-4}$, $K_{a2} = 5.0 \times 10^{-9}$

____

12. In a solution of H_2SO_3 the sulfite concentration is 0.20 M and the pH is 6.50. What is the concentration of H_2SO_3 in the solution?

____

New Terms

none

14.4 BUFFERS

Objectives

To learn how the pH of a solution can be controlled by a mixture of a weak acid or base and its salt. You should learn how to calculate the relative amounts of acid (or base) and salt needed to give a desired pH. You should also be able to calculate the effect on pH produced by addition of small amounts of strong acid or base to a buffer.

Review

Remember, a buffer is a mixture of a weak acid and a weak base. This can be obtained by mixing a weak acid with one of its salts, since the anion of the acid is a weak base. A buffer can also be made by mixing a weak base with one of its salts. The H^+ concentration in an acid buffer can be calculated from the equilibrium constant expression,

$$[H^+] = K_a \frac{[HA]}{[A^-]}$$

Similarly, the OH^- concentration in a basic buffer can be calculated by solving the equilibrium constant expression for $[OH^-]$,

$$[OH^-] = K_b \frac{[B]}{[HB^+]}$$

Review Example 14.11 (which shows you how to determine the acid-to-salt ratio needed to give a desired pH) and Example 14.12 (which shows you how to calculate the effect of additions of strong acid or base to a buffer).

Remember that when a salt like NH_4Cl is dissolved in water it is completely dissociated. If the concentration of NH_4Cl in a solution is 0.10 M, the concentration of NH_4^+ (from this salt) is also 0.10 M. If the equilibrium involves NH_3 and NH_4^+, the Cl^- doesn't matter. It is there only to keep the solution electrically neutral; almost any anion would do. The Cl^- is a spectator ion.

Self-Test

13. Calculate the pH of the following solutions:

(a) 0.25 M $HC_2H_3O_2$, 0.15 M $NaC_2H_3O_2$ ____________

(b) 0.25 M NH_3, 0.15 M NH_4Cl ____________

14. What ratio of formate ion (CHO_2^-) to formic acid ($HCHO_2$) must be maintained to give a solution with a pH = 3.50 (for $HCHO_2$, K_a = 1.8×10^{-4})? ____________

15. How much will the pH change if 0.2 mole of HCl is added to 1.0 liter of a buffer composed of 1 M $HCHO_2$ and 1 M $NaCHO_2$ (the K_a is given in Question 14 above)? ____________

New Terms

buffer

14.5 HYDROLYSIS

Objectives

To consider the equilibria that result when salts of weak acids or bases are dissolved in water.

Review

Hydrolysis consists of the reaction of a substance with water. Several possibilities are dealt with in this section:

(1) Salts of a weak acid and a strong base - An example is $NaC_2H_3O_2$. Only the anion hydrolyzes.

$$A^- + H_2O \rightleftharpoons HA + OH^-$$

The resulting solution is slightly basic. Remember that $K_h = K_w/K_a$.

ON CARD

(2) Salts of a strong acid and a weak base - An example is NH_4Cl. Only the cation hydrolyzes.

$$HB^+ + H_2O \rightleftharpoons B + H_3O^+$$

The resulting solution is slightly acidic. Remember that $K_h = K_w/K_b$.

Numerical problems dealing with hydrolysis equilibria of these two types are illustrated in Examples 14.19 and 14.20 in the text. Compare the method of solution of these examples with that of Example 14.1 in the study guide. Except for the evaluation of the hydrolysis constant, they are really the same. We end up with an expression,

$$\frac{(x)(x)}{(\text{concentration})} = (\text{equilibrium constant})$$

(3) Salts of a weak acid and a weak base - An example is $NH_4C_2H_3O_2$. In this case both the cation and anion hydrolyze. If you have K_a and K_b for the weak acid and base, you can calculate K_h for each of them. If the K_h for the anion is larger than K_h for the cation, the solution is basic. An acidic solution results if K_h for the cation is greater than K_h for the anion.

(4) Salts of polyprotic acids - An example is Na_2CO_3. Remember that you only have to consider the first step in the hydrolysis. If the anion is from a diprotic acid, remember that K_h is calculated using K_{a2} for the second step in the dissociation of the acid. Review Example 14.21.

Self-Test

16. Calculate the H^+ concentration in 0.10 M NaF. For HF, $K_a = 6.5 \times 10^{-4}$. __________

17. Calculate the H^+ concentration in 0.10 M CH_3NH_3Cl. For CH_3NH_2, $K_b = 3.7 \times 10^{-4}$. __________

18. Will a solution of NH_4NO_2 be acidic or basic? $K_a = 4.5 \times 10^{-4}$ for HNO_2, $K_b = 1.8 \times 10^{-5}$ for NH_3. __________

19. Calculate the H^+ concentration in a 0.050 M solution of K_2SO_3. For H_2SO_3, $K_{a1} = 1.5 \times 10^{-2}$, $K_{a2} = 1.0 \times 10^{-7}$. __________

New Terms

hydrolysis
hydrolysis constant

14.6 ACID-BASE TITRATION: THE EQUIVALENCE POINT

Objectives

To see how the pH at the equivalence point in an acid-base titration is influenced by hydrolysis. You should be able to calculate the pH at various points during a titration.

Review

Three situations are discussed in this section:

(1) Titrations of strong acid-strong base. When base is added to the acid complete neutralization occurs and the H^+ concentration is decreased in direct proportion to the amount of OH^- added.

(2) Titration of a weak acid-strong base. At the start of the titration the pH is controlled by the presence of the weak acid. Example 14.1 in the Study Guide illustrated how the H^+ concentration could be obtained for a solution of a weak acid.

After some base has been added, the amount of weak acid has decreased and some of the anion of the acid is generated. In Section 14.4 you saw that a buffer can be prepared as a mixture of a weak acid and its anion. Therefore, between the start of the titration and the equivalence point the pH is calculated in the same way you would compute the pH of a buffer solution.

At the equivalence point the weak acid has been completely neutralized; that is, it has been converted to a salt of the acid. We have just seen in the last section that the anion of a weak acid hydrolyzes. As a result, calculation of the pH at the equivalence point is a hydrolysis problem.

(3) Titration of a weak base-strong acid. The computations involved here are essentially identical to those for the weak acid-strong base.

Self-Test

20. Consider the neutralization of 0.10 M HOCl by the addition of solid NaOH (so that no volume change occurs). Calculate:

(a) the pH before any NaOH is added ________

(b) the pH when half the HOCl has been neutralized ________

(c) the pH at the equivalence point ________
(For HOCl, $K_a = 3.1 \times 10^{-8}$ M)

New Terms

none

14.7 ACID-BASE INDICATORS

Objectives

To learn how an acid-base indicator functions.

Review

An indicator is a weak acid (or base) which possesses one form (HIn) that has a characteristic color different from another form (In^-). The color that is observed in a solution of the indicator is controlled by the ratio of [HIn] to [In^-], which is determined in turn by the H^+ concentration in the solution.

Self-Test

21. The indicator, methyl orange, has pK = 3.5. In acid the indicator is red; in base it is yellow-orange. What color will a solution of methyl orange be if the [H^+] = 2.1 x 10^{-3} M? ________________

New Terms

none

ANSWERS TO SELF-TEST QUESTIONS

1. (a) 1.0 x 10^{-6} (b) 2.4 x 10^{-3} M (c) 1.2 x 10^{-13} M 2. (a) 1.0 x 10^{-10} M (b) 3.6 x 10^{-6} M (c) 1.5 x 10^{-3} M 3. (a) pH = 5.70, pOH = 8.30 (b) pH = 8.07, pOH = 5.93 (c) pOH = 5.47, pH = 8.53 4. (a) 3.2 x 10^{-9} M (b) 4.6 x 10^{-14} M (c) 2.2 x 10^{-1} M 5. 1.4 x 10^{-5} 6. 4.4 x 10^{-4} (did you remember to obtain [OH^-] from pOH?) 7. [OH^-] = 1.9 x 10^{-6} M, [H^+] = 5.3 x 10^{-9} M 8. 4.4 x 10^{-3}% 9. pK_a = 9.31 10. pH = 4.04 11. [H^+] = [HA^-] = 4.5 x 10^{-3} M, [A^{2-}] = K_{a2} = 5.0 x 10^{-9} M 12. 1.3 x 10^{-5} M 13. (a) 4.52 (b) 9.48 14. [CHO_2^-]/[$HCHO_2$] = 0.56 15. pH decreases by 0.17 16. [OH^-] = 1.2 x 10^{-6} M, [H^+] = 8.1 x 10^{-9} M 17. [H^+] = 1.6 x 10^{-6} M 18. K_h for NO_2^- = 2.2 x 10^{-11}, K_h for NH_4^+ = 5.6 x 10^{-10}; the solution will be acidic 19. [OH^-] = 7.1 x 10^{-5} M, [H^+] = 1.4 x 10^{-10} M 20. (a) 4.25 (b) pH = pK_a = 7.51 (c) 10.25 21. red, since [HIn] > [In^-]

15 SOLUBILITY AND COMPLEX ION EQUILIBRIA

This brief chapter concludes the discussion on the quantitative aspects of ionic equilibria by considering equilibria involving salts of low solubility. We will also examine equilibria of complex ions and the way that solubility is affected by complex ion formation.

15.1 SOLUBILITY PRODUCT

Objectives

To deal quantitatively with the solubility equilibria involving "insoluble" salts. You should learn how to write the equilibrium constant expression for the solubility equilibrium. You should be able to calculate K_{sp} from solubility, and solubility from K_{sp}. You should learn how to use the K_{sp} expression to predict whether or not a precipitate will form in a given solution.

Review

Remember that salts are completely dissociated when dissolved in water. The K_{sp} expression involves only the product of ion concentrations raised to exponents that are the coefficients in the chemical equation for the equilibrium.

Problems dealing with K_{sp} can be divided into three classes:

(1) Calculation of K_{sp} from solubility - This is shown in Example 15.1 in the text. Calculate the ion concentrations from the molar solubility and the balanced equilibrium equation. For example, the molar solubility

of Ag_2CrO_4 is 7.8×10^{-5} M. The equilibrium is

$$Ag_2CrO_4 \rightleftharpoons 2Ag^+ + CrO_4^{2-}$$

From the stoichiometry of the equation we would conclude that if 7.8×10^{-5} mole/liter of Ag_2CrO_4 dissolves; then $[Ag^+] = 2(7.8 \times 10^{-5}\ M) = 1.6 \times 10^{-4}$ M and $[CrO_4^{2-}] = 7.8 \times 10^{-5}$ M.

(2) Calculation of solubility from K_{sp} - This is illustrated in Examples 15.3 and 15.4. The important thing to remember with this kind of problem is to work through it methodically. Don't try to skip any of the reasoning steps. If you do, you are likely to make mistakes. Review the reasoning in these examples to be sure you understand it. Note that in constructing the concentration table for the problem the coefficient preceding the x's in the table are the same as the coefficients in the balanced equation for the equilibrium.

(3) Determining whether precipitation will occur - Determine the concentrations of the ions in the solution in question. Then remember the following:

ion product $< K_{sp}$ ion product $= K_{sp}$	no precipitate will form
ion product $> K_{sp}$	precipitate will form

Keep in mind that if the final solution is formed by mixing two solutions, you must consider dilution. Each solute is diluted when the other solution is added. Review Examples 15.5 and 15.6 in the text.

In Example 15.6 note that we can control the concentration of the precipitating ion (S^{2-} in this example), which comes from a weak acid, by properly adjusting the pH of the solution. This example is interesting because it demonstrates how ions can be separated by selective precipitation where the concentration of the precipitating agent (S^{2-} in this case) is controlled by the pH of the solution.

Self-Test

1. The molar solubility of CuCl is 5.7×10^{-4} M. What is the value of K_{sp} for CuCl? __________

2. The molar solubility of $PbBr_2$ is 1.05×10^{-1} M. What is the value of K_{sp}? __________

3. $K_{sp} = 7.0 \times 10^{-10}$ for $SrCO_3$. What is the molar solubility of $SrCO_3$? __________

4. $K_{sp} = 2 \times 10^{-33}$ for $Al(OH)_3$. What is the Al^{3+} concentration in a saturated solution of $Al(OH)_3$?

____

5. A solution is prepared by mixing 100 ml of 0.10 M LiCl solution with 200 ml of 0.30 M NaF solution. LiF has $K_{sp} = 5 \times 10^{-3}$. Will a precipitate form in this solution?

____

6. A solution of 0.010 M Ca^{2+} also contains 0.10 M HF. The solution has had its acidity adjusted to a pH = 2.00 by addition of HCl. Given that for CaF_2, $K_{sp} = 1.7 \times 10^{-10}$ and for HF, $K_a = 6.5 \times 10^{-4}$, will a precipitate of CaF_2 form in this solution?

____

New Terms

solubility product constant

15.2 COMMON ION EFFECT AND SOLUBILITY

Objectives

To see how the solubility of a substance is decreased by the presence of salts that provide a "common ion".

Review

The addition of a common ion to a solution containing a salt in equilibrium with its ions decreases the solubility of the salt. Review Examples 15.7 and 15.8 in the text. Once again, approach this kind of problem in a deliberate, stepwise fashion; don't try to rush or skip steps in the reasoning.

Example 15.8 is typical of the kind of question that so often makes students ask, "When am I supposed to double something and when don't I double something?" Examine the concentration table constructed for this problem. Notice that in the column labeled "Initial concentration" the concentrations of the OH^- from the NaOH, which is already dissolved in the solution, is entered without reference to the equation for the equilibrium. If a solution contains 0.10 M NaOH, the OH^- concentration from this source is 0.10 M. We have not doubled it. Now look at the column headed "Change". This is the column that contains x; the x's have coefficients corresponding to the coefficients of the ions in the balanced equation for the equilibrium. Remember, the only thing ever doubled (or tripled, etc.) are entries in the "Change" column.

Self-Test

7. $K_{sp} = 2 \times 10^{-8}$ for $PbSO_4$. What is the solubility of $PbSO_4$ in 0.010 M Na_2SO_4? __________

8. $K_{sp} = 2 \times 10^{-15}$ for $Fe(OH)_2$. What is the solubility of $Fe(OH)_2$ in a solution having a pH of 10.0? __________

9. $K_{sp} = 1.2 \times 10^{-23}$ for $Zn(OH)_2$. What is the solubility of $Zn(OH)_2$ in 1.0×10^{-3} M $Ba(OH)_2$? __________

10. What is the solubility of $Zn(OH)_2$ in 1.0×10^{-3} M $Zn(NO_3)_2$? __________

New Terms

common ion effect

15.3 COMPLEX IONS

Objectives

To learn what complex ions are and to examine their equilibria. You should learn how to write expressions for the formation constant and the instability constant for a complex ion.

Review

The ions or molecules that attach themselves to the central atom in a complex ion are called ligands. The formation constant (or stability constant) is the K for the reaction in which the complex ion appears as a product (i.e., a formation reaction). For example,

$$Cu^{2+} + 4Cl^- \rightleftharpoons CuCl_4^{2-}$$

$$K_{form} = \frac{[CuCl_4^{2-}]}{[Cu^{2+}][Cl^-]^4}$$

The instability constant is the reciprocal of K_{form}.

New Terms

complex ion
ligand
instability constant
formation constant
stability constant

15.4 COMPLEX IONS AND SOLUBILITY

Objectives

To see how the formation of complex ions can affect solubility.

Review

The formation of complex ions affects the solubility of some salts through a system of simultaneous equilibria, where shifting the position of one equilibrium affects the concentration of a species that is also involved in another equilibrium. This then shifts the position of the second equilibrium. The overall equilibrium constant for the reaction, K_{eq}, is the product of the K_{sp} for the insoluble salt and K_{form} of the complex ion.

$$K_{eq} = K_{sp} \cdot K_{form}$$

Review the sample calculations in Examples 15.9 and 15.10.

Self-Test

11. How much $Cu(OH)_2$ will dissolve in 1.0 liter of 1.0 M NH_3? K_{sp} = 1.6×10^{-19} for $Cu(OH)_2$; $K_{inst} = 2.1 \times 10^{-13}$ for $Cu(NH_3)_4^{2+}$. Ignore the ionization of NH_3 as a weak base.

____

New Terms

none

ANSWERS TO SELF-TEST QUESTIONS

1. 3.2×10^{-7} 2. 4.63×10^{-3} 3. 2.6×10^{-5} 4. 2.9×10^{-9} M 5. yes; $[Li^+][F^-] = (3.3 \times 10^{-2} M)(2.0 \times 10^{-1} M) = 6.6 \times 10^{-3} > K_{sp}$ 6. yes; $[F^-] = 6.5 \times 10^{-3}$, ion product $= 4.2 \times 10^{-7} > K_{sp}$ 7. 2×10^{-6} M 8. 2×10^{-7} M 9. 3.0×10^{-18} M 10. 5.5×10^{-11} M 11. 5.7×10^{-3} mole

16
ELECTROCHEMISTRY

This chapter examines the way redox reactions can be caused to occur by the action of electricity, and the way electricity can be obtained from redox reactions that occur spontaneously. The study of electrochemical properties has wide-range applications, from the construction of new and improved batteries and fuel cells to the study of the electrolyte balance in living cells.

Several new terms are introduced in the introduction to this chapter. They are: electrolytic cell, galvanic cell, voltaic cell, electrolysis.

16.1 METALLIC AND ELECTROLYTIC CONDUCTION

Objectives

To compare the way charge is transported in metals and in solutions of electrolytes.

Review

In metals, conduction occurs by the movement of electrons; in solutions of electrolytes conduction takes place by the movement of ions. Remember that in every microscopic portion of the solution electrical neutrality is maintained, as illustrated in Figure 16.2.

New Terms

metallic conduction
electrolytic conduction

16.2 ELECTROLYSIS

Objectives

To examine the processes that take place at the cathode and anode in an electrolytic cell. You should learn from the examples given how the net reaction in the electrolysis of aqueous solutions is controlled by which redox reactions occur most easily.

Review

In any electrochemical cell, we define the electrodes as:

cathode - electrode where reduction occurs
anode - electrode where oxidation occurs

The net reaction in a cell is the sum of cathode (reduction) and anode (oxidation) half-reactions. The sum is taken so that an equal number of electrons are gained and lost. This is the same procedure you learned in the ion-electron method of balancing equations.

During the electrolysis of aqueous solutions there are usually competing reactions. The half-reactions that occur are those that take place most easily. Review the discussions on the electrolysis of aqueous NaCl, $CuSO_4$ and Na_2SO_4.

Self-Test

1. What is the role of Na_2SO_4 in the electrolysis of aqueous Na_2SO_4 solutions? ______________________

2. Write the anode and cathode reactions for the electrolysis of H_2O.

   ______________________ ______________________

3. In the electrolysis of brine, aqueous NaCl, Cl_2 is produced at the anode rather than O_2. What does this imply about the ease of oxidation of H_2O? ______________________

New Terms

cathode
anode
cell reaction

16.3 PRACTICAL APPLICATIONS OF ELECTROLYSIS

Objectives

To see how electrolysis is applied in industrial processes that affect the way we live.

Review

You should become familiar with the processes involved in the commercial preparation of aluminum and magnesium, and in the purification (refining) of copper.

Self-Test

4. What is the function of the cryolite in the Hall process?

____

5. Write chemical equations for the separation of magnesium from sea water.

____

____

____

6. What reaction occurs at the cathode in the electrolytic refining of copper? _______________________________

New Terms

Hall process
cryolite
electroplating

16.4 QUANTITATIVE ASPECTS OF ELECTROLYSIS

Objectives

To examine the quantitative relationships between the quantity of electricity consumed and the amount of chemical change produced. You should be able to perform calculations of the type illustrated in this section.

Review

The important relationships used in this section are:

$$1 \text{ faraday } (\mathscr{F}) = 96{,}500 \text{ coulombs (coul)}$$
$$1 \text{ coulomb} = 1 \text{ amp sec}$$

The amp sec is obtained as a product of electric current (amp) times time (sec).

A coulometer is a device in which the amount of chemical change produced in one cell is used to determine the number of faradays that have passed through another cell connected to it in series. Be sure to review Examples 16.1 to 16.3. Remember that you must have a balanced half-reaction to solve these problems.

Self-Test

7. How many (a) coulombs and (b) faradays are supplied by a current of 10 amps for 8 hours? (a) ______________ (b) ______________

8. How many grams of Cr will be produced by reduction of Cr^{3+} with a current of 1.50 amps for 30 minutes? ______________

9. How many hours must a current of 14 amps flow to reduce 1.00 mole of Al^{3+} to metallic aluminum in the Hall process? ______________

10. How many grams of Al are deposited on an electrode when 14.0 g of Ag are also produced when the two cells are connected in series? ______________

11. What current would be necessary to oxidize 1.00 g of water in 2 hours? ______________

New Terms

faraday
coulomb
Faraday's law

16.5 GALVANIC CELLS

Objectives

To see how electricity can be provided by a spontaneous redox reaction if the oxidation and reduction half-reactions can be physically separated.

Review

Remember that the redox reactions in a galvanic cell are separated so that the electron transfer occurs through an external circuit. The cell compartments must be connected by a salt bridge or porous partition so that electrical neutrality can be maintained.

In a galvanic cell the anode (where oxidation occurs) is negative; the cathode is positive.

New Terms

galvanic cell
salt bridge

16.6 CELL POTENTIALS

Objectives

To define a quantity which is a measure of the driving force of the redox reaction in a galvanic cell.

Review

The force with which a galvanic cell tends to push electrons through an external circuit is called its emf (electromotive force) and is measured in volts (V). It is also called the cell potential. Standard cell potentials, $\mathcal{E}^{\circ}$, are used when all species are at unit activity and the temperature is 25°C. Cell potentials should be measured with a potentiometer or other similar device which does not draw any current from the cell while the measurement is being made.

Remember that the volt is a measure of the energy that is delivered by a flowing current.

$$1 \text{ volt} = 1 \text{ joule/coul}$$

Self-Test

12. What three factors influence the cell potential for a redox reaction?

____

13. What is observed if one attempts to measure a cell potential with an ordinary voltmeter instead of a potentiometer?

____

14. How much work is done by the flow of 1.20 amp for 5.00 min under a potential of 110 V?

____

New Terms

electromotive force	cell potential
emf	potentiometer
volt	

16.7 REDUCTION POTENTIALS

Objectives

To treat the observed cell potential as the difference between the potentials of competing reduction reactions. Also, to devise a system for tabulating standard reduction potentials for a series of half-reactions. You should learn how to use standard reduction potentials to calculate the cell potential for an overall reaction. You should also learn how to predict the spontaneity of a reaction.

Review

In this section the idea is developed that the measured cell potential can be viewed as the difference between two reduction potentials. The reduction potential is a measure of the tendency of a reduction half-reaction to occur. When the reaction occurs at 25°C with all species at unit concentration, the term standard reduction potential is used.

The standard hydrogen electrode is assigned a reduction potential of exactly 0.000 volts. Other reduction potentials are compared to that of the hydrogen electrode. A positive $\mathcal{E}^0$ means a half-reaction has a greater tendency to occur than the reaction,

$$2H^+ + 2e^- \rightleftharpoons H_2(g)$$

For a given overall reaction that can be divided into half-reactions, the standard cell potential is obtained as,

$$\mathcal{E}^o_{cell} = (\mathcal{E}^o_{substance\ reduced}) - (\mathcal{E}^o_{substance\ oxidized})$$

Review the list of uses to which the table of reduction potentials (Table 16.1) can be put. Remember that a spontaneous reaction occurs only if the calculated $\mathcal{E}^o_{cell}$ is positive. Learn the diagonal relationship described under 3 on Page 466.

Self-Test

15. Calculate the value of $\mathcal{E}^o_{cell}$ for the following reactions.

(a) $Sn^{2+} + H_2 + 2\,OH^- \longrightarrow 2H_2O + Sn$ __________

(b) $4Fe + 3\,O_2 + 12H^+ \longrightarrow 6H_2O + 4Fe^{3+}$ __________

(c) $2Al + 3Zn^{2+} \longrightarrow 2Al^{3+} + 3Zn$ __________

16. Determine whether the following are spontaneous reactions.

(a) $Sn^{2+} + 2SO_4^{2-} \longrightarrow Sn + S_2O_8^{2-}$ __________

(b) $Pb + PbO_2 + 4H^+ + 2SO_4^{2-} \longrightarrow 2PbSO_4 + 2H_2O$

____

(c) $Mn^{2+} + 2Cl^- \longrightarrow Cl_2 + Mn$ __________

17. What is the cell potential and the spontaneous reaction that occurs between the following two half-reactions?

$2NO + 2H^+ + 2e^- \rightleftharpoons H_2N_2O_2$ $\quad \mathcal{E}^o = +0.71\ V$

$Rh^{3+} + 3e^- \rightleftharpoons Rh$ $\quad \mathcal{E}^o = +0.80\ V$

____

New Terms

reduction potential
standard reduction potential
standard hydrogen electrode
oxidation potential

16.8 SPONTANEITY OF OXIDATION-REDUCTION REACTIONS

Objectives

To relate the standard potential for a cell to ΔG^o for the cell reaction.

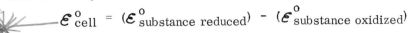

Review

Remember the equation,

$$\Delta G^{o} = -n\mathcal{F}\mathcal{E}^{o}$$

where n is the number of moles of electrons transferred and $\mathcal{F}$ is the faraday constant.

Self-Test

18. The standard cell potential for the reaction,

$$Mg + Ni^{2+} \longrightarrow Ni + Mg^{2+}$$

is 2.13 V. Calculate ΔG^{o} for the reaction. ____________

19. What is ΔG^{o} for the reaction in Self-Test Question 13? ____________

New Terms

none

16.9 THERMODYNAMIC EQUILIBRIUM CONSTANTS

Objectives

To relate the standard cell potential to the thermodynamic equilibrium constant.

Review

Learn the equation,

$$\mathcal{E}^{o} = \frac{0.0592}{n} \log K_c$$

This equation can be used to evaluate K_c from measured cell potentials. Review Example 16.5 in the text.

Self-Test

20. Using the data in Table 16.1, evaluate the thermodynamic equilibrium constant for the reaction, $2Fe^{2+} + Br_2 \rightleftharpoons 2Fe^{3+} + 2Br^{-}$

____

21. What is the equilibrium constant for the reaction in Self-Test Question 17? ____________

New Terms

none

16.10 CONCENTRATION EFFECT ON CELL POTENTIAL

Objectives

To obtain a relationship between the cell potential and the concentrations of the species involved in the cell reaction. You should learn how a concentration cell operates and how to calculate its emf. You should also learn that the emf of a concentration cell can be used to determine concentrations.

Review

Learn the Nernst equation,

$$\mathcal{E} = \mathcal{E}^{\circ} - \frac{0.0592}{n} \log Q$$

where Q represents the mass action expression for the cell reaction (omitting the concentrations of any pure solids). Review the principles of the concentration cell on Page 471. Remember that

$$\mathcal{E}_{cell} = -\frac{0.0592}{n} \log \frac{[M^{n+}]_{dil}}{[M^{n+}]_{conc}}$$

On Pages 473 to 475 several applications of the Nernst equation are shown. Review these and try the questions in the following Self-Test.

Self-Test

22. Calculate the potential of the concentration cell consisting of one iron electrode immersed in a 0.0010 M Fe^{3+} solution and another iron electrode immersed in a 0.10 M Fe^{3+} solution. ____________

23. Calculate the potential generated by the cell reaction,
$2Al + 3Fe^{2+}$ (0.0010 M) $\longrightarrow 2Al^{3+}$ (0.10 M) $+ 3Fe$ ____________

24. A cell is constructed in which one electrode consists of a Zn electrode dipping into 1.0 M $ZnSO_4$. The other electrode consists of a silver electrode in a solution containing Ag^+ of unknown concentration. The cell potential is observed to be 1.40 V with the Zn serving as the anode.

(a) What is the cell reaction? ______________________

(b) What is the Ag^+ concentration? ______________________

New Terms

Nernst equation
concentration cell

16.11 ION-SELECTIVE ELECTRODES

Objectives

To describe some applications of galvanic cells that respond to changes in the concentration of a single ion.

Review

This section is to provide you with an awareness of the scope of applications of ion-selective electrodes.

New Terms

ion-selective electrode
glass electrode
enzyme-substrate electrode

16.12 SOME PRACTICAL GALVANIC CELLS

Objectives

To examine the chemistry of common (and not so common) practical galvanic cells that are used as a source of electricity.

Review

Learn the chemical reactions that occur at the anode and cathode in the dry cell, the lead storage battery, and the nickel-cadmium cell. Review the principles of operation of the fuel cell as well as its advantages over conventional power sources.

Self-Test

25. After having reviewed this last section, write the chemical equations for the cathode and anode reactions in the dry cell, lead storage battery, and nickel cadmium cell.

New Terms

dry cell
lead storage battery
nickel-cadmium cell
fuel cell

ANSWERS TO SELF-TEST QUESTIONS

1. It maintains electrical neutrality. 2. anode: $2H_2O \longrightarrow O_2 + 4H^+ + 4e^-$ cathode: $2H_2O + 2e^- \longrightarrow H_2 + 2\ OH^-$ 3. H_2O is more difficult to oxidize than Cl^-. 4. It lowers the melting point of Al_2O_3.
5. $Mg^{2+} + 2\ OH^- \longrightarrow Mg(OH)_2(s)$
$Mg(OH)_2 + 2HCl \longrightarrow MgCl_2 + 2H_2O$
$MgCl_2(l) \longrightarrow Mg(l) + Cl_2(g)$
6. $Cu^{2+}(aq) + 2e^- \longrightarrow Cu(s)$ 7. (a) 2.88×10^5 coul (b) 2.98 $\mathcal{F}$
8. 0.485 g 9. 5.74 hr 10. 1.17 g Al 11. 1.49 amp 12. nature of the species involved, their concentration, temperature 13. The voltage is lower when measured with a voltmeter. 14. 1.20 amp x 5.00 min x 60 sec/min = 360 coul; 360 coul x 110 J/coul = 39.6 kJ 15. (a) 0.69 V (b) 1.27 V (c) 0.91 V 16. (a) not spont. (b) spont. (c) not spont. 17. $\mathcal{E}^\circ = 0.09$ V; $2Rh^{3+} + 3H_2N_2O_2 \longrightarrow 6NO + 6H^+ + 2Rh$ 18. -411 kJ (-98.2 kcal)
19. -17 kJ (-4 kcal) 20. 6.5×10^{10} 21. 1.3×10^9 22. 0.039 V
23. 1.20 V 24. (a) $Zn + 2Ag^+ \longrightarrow Zn^{2+} + 2Ag$ (b) $[Ag^+] = 10^{-3}$ M
25. see Pages 477-480 in the text.

17 COVALENT BONDING AND MOLECULAR STRUCTURE

This chapter describes the modern theories of chemical bonding and how they can account for (or predict, in some cases) molecular structure. The thing to keep in mind throughout discussions of the different approaches is that each theory is attempting to describe the same thing and each, in its own way, succeeds to a degree. The theories, then, present alternative views of the same phenomenon.

17.1 VALENCE BOND THEORY

Objectives

To describe how this theory views the formation of a chemical bond. You should learn the basic ideas on which the theory is based.

Review

This theory says that a bond is formed by sharing a pair of electrons between overlapping atomic orbitals. Only two electrons can be shared between two orbitals. When orbitals come together, they must each have one electron, or one must be empty if the other orbital supplies two electrons (a coordinate covalent bond). When p orbitals are used in bonding, bond angles tend toward 90°.

Self-Test

1. What would you predict for the structure (shape and bond angles) of:

   (a) H_2S ____

(b) PH_3 ______________________________

New Terms

valence bond theory

17.2 HYBRID ORBITALS

Objectives

To see how atomic orbitals on an atom can mix to form new hybrid orbitals that possess new directional properties. You should learn how molecular shapes can be accounted for by the use of hybrid orbitals. You should learn the directional properties of different kinds of hybrid orbitals.

Review

Learn the orientations (summarized in Figure 17.7) of the different hybrid orbital sets in Table 17.1. In applying the information in this table to describing molecular structure, remember that when a central atom enters into bonding with several others, it must supply one unpaired electron for each of the atoms to which it is bonding. Review the descriptions for BeH_2, CH_4 and SF_6. Note that additional electron pairs belonging to the central atom which are not used in bonding can reside in hybrid orbitals too. This is described for H_2O and NH_3. Remember that when hybrids are formed, s and p orbitals are used first, followed by d orbitals.

Self-Test

2. What kind of hybrid orbitals are used by the central atom in each of the following? What molecular structure is expected?

   (a) $SiCl_4$ ______________ ______________________

   (b) BCl_3 ______________ ______________________

   (c) SF_4 ______________ ______________________

   (d) PCl_5 ______________ ______________________

   (e) $SnCl_2$ ______________ ______________________

New Terms

hybrid orbitals

17.3 MULTIPLE BONDS

Objectives

To see how multiple bonds are described in terms of orbital overlap.

Review

Multiple bonds are usually formed when unpaired electrons would occur on atoms if the rules in the last section were followed. In ethylene, for example, each carbon atom is bonded to three other atoms.

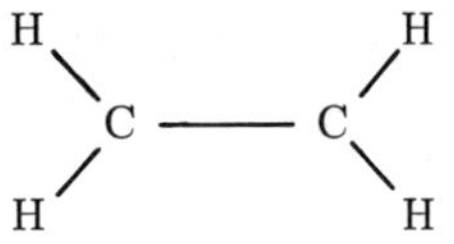

This requires three unpaired electrons on the carbon atoms and if no double bond were formed, we would have the situation,

C ↑ ↑x ↑x ↑x (x's are electrons from other atoms)
sp^3

This leaves an unpaired electron in an sp^3 hybrid orbital. This tends to displease Mother Nature. As a result, in this kind of situation we only use three hybrid orbitals (sp^2) for σ-bonds. The remaining electron remains in an unhybridized orbital and can then pair with another electron on the neighboring carbon atom to form a π-bond.

Lewis structures are based on the valence bond description of bonding. If a Lewis structure has multiple bonds, you should account for the bonding as described in this section. Be sure you understand the meaning of σ-bond and π-bond.

Self-Test

3. Indicate the number of σ- and π-bonds in the bonds of the following molecules:

(a) CO_2 ______________________

(b) CO_3^{2-} ______________________

(c) N_2 ______________________

(d) C_2^{2-} ______________________

(e) GeH_4 ______________________

New Terms

σ-bond
π-bond

17.4 RESONANCE

Objectives

To relate the Lewis structures for molecules that exhibit resonance to the valence bond description of bonding.

Review

The point that is made here is that Lewis structures are, in effect, simplified versions of valence bond structures.

New Terms

none

17.5 MOLECULAR ORBITAL THEORY

Objectives

To understand the concept of how molecular orbitals are created from atomic orbitals. You should learn the difference between bonding and antibonding molecular orbitals and how the electronic structure of a molecule is obtained by filling molecular orbitals. You should also learn how molecular orbital theory avoids the idea of resonance.

Review

Molecular orbitals are obtained by alternately adding and subtracting atomic orbitals that overlap from different atoms. Bonding orbitals concentrate electron density between nuclei, thereby binding the atoms together. Antibonding orbitals place electron density outside the region between nuclei and, when occupied by electrons, decrease the stability of the molecule.

Molecular orbitals are filled following the same rules for filling atomic orbitals in atoms: (1) electrons go into lowest energy orbitals first; (2) no more than two electrons may occupy the same orbital; (3) when there are orbitals of the same energy, the electrons spread out as much as possible with spins in the same direction.

Learn how to calculate the net bond order from the number of bonding and antibonding electrons. You should also learn the energy level diagram for diatomic molecules (Figure 17.17).

Notice that molecular orbital theory avoids resonance by permitting molecular orbitals to spread out over more than two atoms.

Self-Test

4. On a separate piece of paper, sketch the molecular orbital energy level diagram for diatomic molecules having the second shell (2s and 2p subshells) as their valence shell. Use this diagram to answer the following questions.

   (a) Is the molecule Be_2 stable? ________

   (b) Which species is more stable, O_2 or O_2^+? ________

   (c) Which species is more stable, N_2 or N_2^+? ________

   (d) Which has the shorter bond, C_2^- or C_2^{2-}? ________

New Terms

molecular orbital
bonding orbital
antibonding orbital
net bond order

17.6 ELECTRON-PAIR REPULSION THEORY OF MOLECULAR STRUCTURE

Objectives

To learn a very simple way of predicting molecular structure (geometry) by using electron-dot formulas.

Review

The key to success in applying this theory is knowing how to write electron-dot formulas. If you are a little "rusty" on this, review Sections 4.4 and 4.5 in the text and the Study Guide.

It's important that you know the geometrical arrangements that are found for different numbers of electron pairs. This is summarized in

Figure 17.19. Also review the structures that arise from these electron pair arrangements in Figures 17.20 to 17.22. Learn the names that go with these structures (Table 17.2).

As a final note, remember that the term, molecular structure, refers to the location of the atomic nuclei, not the arrangement of the electron pairs. Thus, we say that the NH_3 molecule has a pyramidal structure, even though the electron pairs are arranged tetrahedrally.

Self-Test

5. For each of the following, predict the arrangement of electron pairs and the molecular structure.

	electron arrangement	molecular structure
(a) AsH_3		
(b) $AlCl_4^-$		
(c) BrF_3		
(d) NO_2^-		
(e) ICl_2^-		
(f) H_3O^+		

New Terms

electron pair repulsion theory

ANSWERS TO SELF-TEST QUESTIONS

1. (a) H-S-H angle, 90°; angular molecule (b) H-P-H angles, 90°; pyramid-shaped molecule 2. (a) sp^3, tetrahedral (b) sp^2, triangular (c) dsp^3, molecule with atoms at four of the five corners of a trigonal bipyramid (d) dsp^3, trigonal bipyramid (e) sp^2, angular, with atoms at two of the three corners of a triangle. 3. (a) two σ-bonds, two π-bonds (b) three σ-bonds, one π-bond (c) one σ-bond, two π-bonds (d) one σ-bond, two π-bonds (e) four σ-bonds 4. (a) no, net bond order = zero (b) O_2^+ is more stable since net bond order is greater than for O_2 (c) N_2, larger net bond order (d) C_2^{2-} (same structure as N_2), larger net bond order

5.

	electron arrangement	molecular structure
(a)	tetrahedral	pyramidal
(b)	tetrahedral	tetrahedral
(c)	trigonal bipyramidal	T-shaped
(d)	planar triangular	angular
(e)	trigonal bipyramidal	linear
(f)	tetrahedral	pyramidal

18 CHEMISTRY OF THE REPRESENTATIVE ELEMENTS: PART I, THE METALS

The chemical behavior of the elements is described in this and the next two chapters. Throughout this discussion we will be focusing on learning trends and relationships among physical and chemical properties and structure, rather than simply learning specific details. The details in the text are intended, for the most part, to illustrate trends. Use the Study Guide to help you concentrate on the key points developed in each section.

18.1 METALS, NONMETALS, AND METALLOIDS

Objectives

To describe the classification of the elements into these three categories. You should learn how the elements are divided and the properties of each class.

Review

Below are summarized some properties of these three classes of elements:

Metals: (a) have basic oxides; (b) are good conductors of heat and electricity; (c) exhibit, almost exclusively, positive oxidation states;

(d) usually combine with nonmetals and metalloids.

Nonmetals: (a) have acidic oxides; (b) are insulators (both thermal and electrical); (c) exhibit both positive and negative oxidation states; (d) combine with metals, other nonmetals, and with metalloids.

Metalloids: (a) have oxides that are usually acidic (some are amphoteric); (b) are semiconductors; (c) usually exhibit positive oxidation states except when combined with active metals; (d) combine with both metals and nonmetals.

You should know where these classes of elements are located in the periodic table. You should also try to understand how their general properties are related to electronic structure.

Self-Test

1. From their positions in the periodic table, how would you classify the following elements?

   (a) Ge ________________ (d) Np ________________

   (b) Cl ________________ (e) B ________________

   (c) Ta ________________ (f) Dy ________________

2. Why do metals tend to form a metallic lattice rather than discrete molecules? ____

   ____

New Terms

galvanizing
pickling

18.2 TRENDS IN METALLIC BEHAVIOR

Objectives

To learn how metallic character varies according to an element's position in the periodic table.

Review

Electronegativity serves as a guide to metallic character; as electronegativity increases, metallic character decreases. Learn the way electronegativity varies throughout the periodic table and use this to guide you in your estimate of relative degrees of metallic behavior.

Self-Test

3. Choose the most metallic element in each of the following sets:

   (a) Pb Sn Bi Sb ________

   (b) Be Mg Al B ________

   (c) Si As Ge P ________

   (d) Sn Ge Sb As ________

4. Which of the following oxides would you expect to be most basic?

   (a) MgO or Al_2O_3 ________ (b) B_2O_3 or Al_2O_3 ________

New Terms

none

18.3 PREPARATION OF METALS

Objectives

To learn what kinds of methods can be used to extract metals from their compounds.

Review

There are three methods described in this section:

(1) Thermal decomposition: This requires that the metal compound being decomposed have a small ΔH_f; otherwise excessively high temperatures must be reached to produce a measurable amount of the free metal. Be sure you understand the thermodynamic argument presented for this in the text. Review Examples 18.1 and 18.2 in the text.

(2) Reduction using a chemical reducing agent: This can only be used economically on metals of moderate activity. Two important chemical reducing agents are carbon and hydrogen.

(3) Electrolytic reduction: This must be used for very active metals. Halide salts are usually chosen for electrolysis because of their relatively low melting points.

Self-Test

5. What types of processes would probably be used to extract the metals from the following compounds?

   (a) PbO ____________________

   (b) HgO ____________________

   (c) NaCl ____________________

New Terms

none

18.4 CHEMICAL PROPERTIES AND TYPICAL COMPOUNDS

Objectives

To survey the typical properties and compounds exhibited by the representative metals.

Review

Below are the major points to review in this section:

(1) General properties. The compounds of the metals in Groups IA and IIA show generally remarkable similarities in properties within their respective groups.

(2) Behavior toward O_2. The Group IA metals show unusual behavior toward molecular oxygen. Learn the compounds formed when they react with O_2. What is a peroxide? What is a superoxide? What is produced when they react with water?

(3) Diagonal relationship. Learn what is meant by the term "diagonal relationship". What elements show this relationship? How is the ionic potential involved? What are some of the similarities (chemical and structural) between Be and Al compounds?

Self-Test

6. What compounds are produced when the following metals react with O_2?

   (a) Li __________ (b) Na __________ (c) Cs __________

7. Complete the following equations:

(a) $O_2^- + H_2O \longrightarrow$ ______________________

(b) $O_2^{2-} + H_2O \longrightarrow$ ______________________

8. What two sets of metals exhibit the diagonal relationship referred to in this section? ______________________

9. Arrange the following ions in order of increasing ionic potential:

	ion	radius (Å)
(a)	Er^{3+}	0.96
(b)	Be^{2+}	0.39
(c)	Ti^{4+}	0.68

____

10. How are the structures of $AlCl_3(s)$ and $BeCl_2(s)$ similar?

____

New Terms

ionic potential
dimeric
diagonal relationship

18.5 OXIDATION STATES

Objectives

To examine the oxidation states that are observed for the representative metals.

Review

Several ideas are presented in this section:

(1) As expected, the oxidation states are controlled by the electron configurations of the metals. This is reviewed for Groups IA to VA.

(2) The standard reduction potentials, $\mathcal{E}^o$, become more negative as one proceeds down within a group. Li has an unusually negative $\mathcal{E}^o$ because of its very large hydration energy.

(3) The metals in Group IIA only show a +2 oxidation state because of the very high hydration energy and lattice energy of the M^{2+} ion, as compared to the M^+ ion.

(4) In Groups IIIA to VA the heavy elements exhibit more than one oxidation state. The lower oxidation state becomes more stable as one proceeds down within a group.

Learn the explanations given to account for these observations.

Self-Test

11. Why doesn't Ca exhibit a +3 oxidation state? ______________________

____

12. Why is Pb^{4+} a better oxidizing agent than Sn^{4+}? ______________

____

13. Why do the heavier elements in Groups IIIA, IVA and VA seem to prefer the lower oxidation state? ______________________

____

New Terms

none

18.6 THE COVALENT/IONIC NATURE OF METAL COMPOUNDS

Objectives

To examine trends in covalent character in metal compounds. You should learn what factors control the relative degree of covalent character in metal-nonmetal bonds. You should be able to use these ideas to make comparisons.

Review

The trends in covalent character examined in this section are related to polarization of ions by those of opposite charge. This pulls some electron density between the two nuclei, in effect producing some degree of covalent bonding between the ions. The following rules apply:

(1) For a given size, multiply charged cations produce a greater degree of polarization of the anion than do singly charged cations.

(2) For ions of a given charge, the polarizing effect becomes greater as the size becomes smaller.

(3) Cations with a pseudonoble gas configuration or cations that immediately follow the transition elements are more effective at polarizing an anion than cations of the same size with a noble gas configuration.

(4) For a given charge, anions become more easily polarized as their size becomes larger.

(5) For a given size, anions become more easily polarized as their charge becomes greater.

Properties that can be related to covalent character are color (the deeper the color, the more covalent the bond) and melting point (for similar structure types, the higher the melting point, the more ionic the bond).

Self-Test

14. Choose the member of each pair below that has the greatest degree of covalent character.

(a) $MgCl_2$ or $BeCl_2$ ____________________

(b) $AlCl_3$ or $AlBr_3$ ____________________

(c) $MgCl_2$ (Mg-Cl distance = 2.46 Å) or $ZnCl_2$ (Zn-Cl distance = 2.42 Å) ____________________

(d) BeF_2 (Be-F distance = 1.6 Å) or SiF_4 (Si-F distance = 1.56 Å) ____________________

(e) Na_2S (Na-S distance = 2.83 Å) or NaCl (Na-Cl distance = 2.81 Å) ____________________

15. Choose the member of each pair that would have the lowest melting point.

(a) MgO or MgS _______________

(b) Al_2S_3 or Ga_2S_3 _______________

16. Choose the member of each pair that would be most deeply colored.

(a) Ag_2O or Ag_2S _______________

(b) SnS or SnSe _______________

New Terms

polarization
charge transfer absorption band

18.7 HYDROLYSIS

Objectives

To see what factors influence the degree of hydrolysis of metal ions.

Review

Metal ions tend to become more extensively hydrolyzed as their size decreases and/or their charge increases.

Self-Test

17. Choose the member of each pair below that is expected to hydrolyze to the greatest degree.

(a) Bi^{3+} or Bi^{5+} ____________

(b) Be^{2+} or Mg^{2+} ____________

New Terms

none

ANSWERS TO SELF-TEST QUESTIONS

1. (a) metalloid (b) nonmetal (c) metal (d) metal (e) metalloid (f) metal 2. They cannot complete their valence shells by electron-sharing (covalent bonds) and they are unable to accept electrons from other atoms of the same type because their electron affinities are so low. The metallic lattice is a compromise. 3. (a) Pb (b) Mg (c) Ge (d) Sn 4. (a) MgO (b) Al_2O_3 5. (a) chemical reduction (b) thermal decomposition (c) electrolysis 6. (a) Li_2O (b) Na_2O_2 (c) CsO_2 7. (a) $OH^- + H_2O_2 + O_2$ (b) $OH^- + H_2O_2$ 8. Li and Mg; Be and Al 9. $Er^{3+} < Ti^{4+} < Be^{2+}$ 10. Both Al and Be are surrounded tetrahedrally by four Cl atoms. 11. The ionization energy is so high that sufficient energy cannot be recovered through the lattice energy or hydration energy. 12. The relative stability of Pb^{2+} over Pb^{4+} is greater than the relative stability of Sn^{2+} over Sn^{4+}. 13. The M-X bond strength decreases with increasing atomic number within a group. 14. (a) $BeCl_2$ (b) $AlBr_3$ (c) $ZnCl_2$ (d) SiF_4 (e) Na_2S 15. (a) MgS (b) Ga_2S_3 16. (a) Ag_2S (b) SnSe 17. (a) Bi^{5+} (b) Be^{2+}

19 CHEMISTRY OF THE REPRESENTATIVE ELEMENTS: PART II, THE METALLOIDS AND NONMETALS

In this chapter we complete our discussion of the descriptive chemistry of the representative elements by examining the structure and properties of the nonmetals and metalloids and their compounds. Once again, you should not attempt to memorize all the detailed information given here. Study and learn the trends that help you remember the specifics.

19.1 THE FREE ELEMENTS

Objectives

To learn how nonmetals and metalloids can be extracted from their compounds.

Review

Metalloids are usually obtained by chemical reduction of one of their compounds.

Some nonmetals exist in nature only in uncombined forms (the noble gases). Others may be found both uncombined and in compounds (O_2, N_2, C, S). Some are only found in combined form (the halogens).

Halogens (except fluorine) can be obtained from their simple salts by chemical oxidation. Fluorine can only be obtained by electrolysis. Review the relative strengths of the free halogens as oxidizing agents.

Self-Test

1. Write the equation for the reduction of Sb_2O_3 with H_2.

   ____

2. Why can't F_2 be obtained by electrolysis of aqueous NaF?

   ____

3. Arrange the halogens in order of decreasing strengths as oxidizing agents.

   ____

New Terms

none

19.2 MOLECULAR STRUCTURE OF THE NONMETALS AND METALLOIDS

Objectives

To look for generalizations that can be applied to understanding the degrees of complexity of the structures of the elemental nonmetals and metalloids.

Review

The key to understanding the molecular structures of these elements is the idea that elements in period 2 are able to form strong π-bonds while elements in the following periods form much weaker π-bonds. As a result, double and triple bonds are observed among period 2 elements, but period 3 elements tend to favor two or three single bonds to separate atoms, rather than π-bonds to just one other atom.

The following are the most important items to review:

(1) the relationships between the structures of elemental chlorine, sulfur, phosphorus, and silicon

(2) the structural changes that occur when sulfur is heated

(3) the structure of graphite which accounts for its electrical conductivity

(4) allotropism involving oxygen

(5) the relationship between ozone and smog

(6) the unusual structure of boron

Self-Test

4. How many atoms do the following elements bond to in their elemental forms?

(a) Cl ________ (c) P ________

(b) S ________ (d) Si ________

5. How does the molecular structure of S change as it is heated?

____

6. How does graphite differ from diamond? ________________

____

7. How is O_3 produced in smog? ____________________

____

New Terms

allotropism
ozone
amorphous sulfur

19.3 OXIDATION NUMBERS

Objectives

To consider oxidation numbers as a possible way of classifying the chemistry of the nonmetals and metalloids.

Review

In general, nonmetals and metalloids exhibit multiple oxidation states. The oxidation state concept is of only limited usefulness in coordinating the chemistry of these elements.

New Terms

none

19.4 NONMETAL HYDRIDES

Objectives

To consider the composition of some of the nonmetal hydrides (binary compounds of the nonmetal with hydrogen).

Review

Ordinarily, hydrogen is only able to form one covalent bond to another atom and this controls the composition of the simple hydrides. Complex hydrides occur when nonmetals are bonded to themselves as well as to hydrogen. Catenation is the term applied to describe the ability of an element to bond to itself (either in hydrides, or elsewhere). The element that exhibits the greatest degree of catenation is carbon.

Self-Test

8. Write the chemical equation for the reaction of sulfur with SO_3^{2-} in aqueous solution. ______________________

9. What are the formulas for the following?

   (a) polysulfide ion ____________

   (b) tetrathionate ion ____________

New Terms

catenation

19.5 PREPARATION OF THE HYDRIDES

Objectives

To outline general methods of preparation of nonmetal hydrides.

Review

Two methods for preparing hydrides are discussed:

(1) direct combination of the elements
(2) reaction of the nonmetal anion X^{n-} with a source of protons

Method 1 is limited, for practical purposes, to nonmetal hydrides with negative ΔG_f^o .

In Method 2 the anion, X^{n-}, is reacted with a Brønsted acid;

$$X^{n-} + nHA \longrightarrow H_nX + nA^-$$

The strength of the acid required is inversely proportional to the basicity of the anion, X^{n-}. Review the trends in basicity of the anions across periods and down groups.

Self-Test

10. Why is it impractical to prepare nonmetal hydrides with positive ΔG_f^o by direct combination of the elements? ______________________

____

11. Which of the following ions react most completely with H_2O?

(a) N^{3-} or P^{3-} ________

(b) P^{3-} or S^{2-} ________

(c) Si^{4-} or C^{4-} ________

(d) Cl^- or P^{3-} ________

New Terms

none

19.6 BORON HYDRIDES

Objectives

To examine some of the unique features of the boron hydrides. You should pay particular attention to the mode of bonding in these compounds.

Review

The most unusual feature of the boron hydrides is the three-center bond - a bond that holds three nuclei together but which contains a single pair of electrons. It is a bond that is spread out (delocalized) over three nuclei.

New Terms

three-center bond

19.7 GEOMETRIC STRUCTURES OF THE NONMETAL HYDRIDES

Objectives

To review the types of structures found for the nonmetal hydrides.

Review

The structures of most simple nonmetal hydrides can be predicted by the electron pair repulsion theory (Chapter 17). The boron hydrides represent a special case.

The structures of the boron hydrides range from approximately tetrahedral arrangements of atoms about boron (in B_2H_6) to more complex structures based on the octahedron and the icosahedron.

Self-Test

12. What are the molecular structures of the following?

(a) PH_3 ________________

(b) SiH_4 ________________

(c) SeH_2 ________________

(d) $B_{12}H_{12}^{2-}$ ________________

New Terms

none

19.8 OXYGEN COMPOUNDS OF THE NONMETALS

Objectives

To review types of oxygen compounds formed by nonmetals and the general relationships between them.

Review

The relationship between the nonmetal oxides and the oxoacids is important to remember. Nonmetal oxides are often able to react with water to produce oxoacids. Nitrogen dioxide undergoes disproportionation in water to give HNO_3 and NO.

New Terms

disproportionation

19.9 PREPARATION OF NONMETAL OXIDES

Objectives

To outline general methods by which the nonmetal oxides may be prepared.

Review

It is pointed out in this section that there are many different kinds of reactions which occur that can produce nonmetal oxides. Three general types of reactions are described:

(1) direct combination of the elements

(2) combustion of nonmetal hydrides to produce the nonmetal oxide and water

(3) redox reactions

Review the general principles involved in each. Also review the chemistry of the industrially important Ostwald process.

Self-Test

13. What would you expect the products of combustion to be when the following are reacted with O_2?

(a) H_2S ________________ (c) C_2H_6 ________________

(b) NH_3 ________________ (d) SiH_4 ________________

New Terms

none

19.10 THE STRUCTURE OF NONMETAL OXIDES

Objectives

To look for generalizations that may be applied to the structure of the nonmetal oxides.

Review

Once again, we find the complexities of structures to be controlled by the relative abilities of period 2 elements and those from the following periods to form π-bonds. The period 2 elements can form simple oxides because oxygen atoms can form multiple bonds to them. With elements from the following periods the oxygen atoms tend to form bonds bridging nonmetal atoms and complex structures result.

Note that sulfur is able to form multiple bonds with oxygen in SO_2 and SO_3.

Examine the structures of P_4O_6 and P_4O_{10}. Compare them with the structure of P_4 (Page 552).

Examine the structure of quartz. Notice that each Si atom is bonded to four oxygens. Also notice that the -O-Si-O- chains form spirals that can twist either clockwise or counterclockwise. This gives two types of quartz crystals.

New Terms

none

19.11 SIMPLE OXOACIDS AND OXOANIONS

Objectives

To examine the chemical characteristics of the simple oxoacids and their anions.

Review

Each of the oxoacids possesses one or more hydroxyl groups that are polarized by the nonmetal atom, thereby permitting the H atom to break away as H^+.

Oxoacids can often be prepared by reaction of the appropriate nonmetal oxide with water. Review the reactions of SO_2, SO_3, P_4O_6, P_4O_{10}, and NO + NO_2 with water. If the oxoacid is a weak acid, it can be prepared by adding a strong acid to a solution containing the oxoanion.

This section also demonstrates the importance of kinetics in controlling the outcome of chemical reactions.

Self-Test

14. What products are formed in the reaction of each of the following with water?

(a) P_4O_6 ____________

(b) NO_2 ____________

(c) NO_2 + NO ____________

(d) SO_3 ____________

(e) CO_2 ____________

15. Why is H_3PO_3 only a diprotic acid? ____________

____

16. Which of the hypohalite ions, OCl^-, OBr^-, or OI^-, should disproportionate to the greatest extent, according to thermodynamics?

____

New Terms

hydroxyl group

19.12 POLYMERIC OXOACIDS AND OXOANIONS

Objectives

To examine the structural relationships that exist among the various polymeric oxoanions (and among their parents, the polymeric oxoacids).

Review

The polymeric oxoacids can be viewed as resulting from the condensation of simpler species by the formation of oxygen bridges. These bridges result from the removal of the constituents of water from two hydroxyl groups on adjacent molecules.

What you should concentrate on in this section is the relationship among the ortho-, pyro- and meta- series of oxoacids and anions. Pay special attention to the summaries provided in Tables 19.8 and 19.9.

Note the chains, sheets and three-dimensional networks formed when two, three and four oxygen bridges, respectively, are formed by SiO_4 tetrahedra.

Self-Test

17. What are the empirical formulas of the oxoanions that we would refer to as:

    (a) ortho ______________ (c) meta ______________

    (b) pyro ______________

18. Why doesn't chlorine form a meta- series compound?

    ____

19. What type of structure occurs when SiO_4 tetrahedra are linked by:

    (a) three oxygen bridges ______________________

    (b) two oxygen bridges ______________________

New Terms

ortho- acids and anions
pyro- acids and anions
meta- acids and anions

19.13 HALOGEN COMPOUNDS OF THE NONMETALS

Objectives

To examine the kinds of binary halogen compounds that are formed by the nonmetals and to study the factors that affect their reactivity toward water.

Review

This section divides the binary halogen compounds into two categories, those possessing four or less electron pairs in the valence shell of the central atom and those in which the central atom exceeds an octet. In the first category, the number of halogens that become attached to the central atom is equal to the number of electrons that the central atom requires to achieve an octet. The boron halides are an exception since boron only has three electrons.

Compounds in the second category are restricted to elements beyond the second period. When the octet is exceeded, d orbitals are employed to create hybrids that can accommodate the required number of electrons.

The relative sizes of the central atom and the halogen also influence the number of a given halogen able to fit about a given central atom (Table 19.10).

The reactivity of the halogen compounds toward water is discussed in terms of the interplay between kinetics and thermodynamics in determining the outcome of reaction. Study the reasons for the differing stabilities toward hydrolysis. Study the mechanism of polymerization of silicones.

Self-Test

20. Which compound is likely to be more unstable, SBr_4 or $SeBr_4$? ______

21. Why don't we observe more than four halogen atoms bonded to a period 2 element? ______

______

22. What reason is given to explain why $SiCl_4$ is susceptible to hydrolysis but CCl_4 is not? ______

______

23. What accounts for the high degree of chemical inertness of SF_6?

____

____

24. If some $(CH_3)SiCl_3$ were added to the $(CH_2)_2SiCl_2$ prior to hydrolysis and polymerization, what effect would this tend to have on the structure of the resulting silicone polymer? _______________

____

New Terms

silicone

19.14 NOBLE GAS COMPOUNDS

Objectives

To examine the kinds of compounds that are formed by the noble gases.

Review

Only the heavier noble gases (Kr, Xe and Rn) have been shown to form compounds. They react with only the most electronegative elements (F and O). Bonding in these compounds involves expansion of the octet, with the participation of d orbitals. Review the bonding in XeF_2 and XeF_4.

New Terms

clathrate

ANSWERS TO SELF-TEST QUESTIONS

1. $Sb_2O_3 + 3H_2 \longrightarrow 2Sb + 3H_2O$ 2. water is more easily oxidized than F^- 3. $F_2 > Cl_2 > Br_2 > I_2$ 4. (a) 1 (b) 2 (c) 3 (d) 4 5. $S_8(s) \longrightarrow S_8(l) \longrightarrow S_x(l) \longrightarrow$ shorter chains of S atoms 6. graphite - planar sheets held to each other by London forces; diamond - three-dimensional covalent structure 7. $NO_2 \xrightarrow{h\nu} NO + O$; $O + O_2 \longrightarrow O_3$
8. $S + S_3^{2-} \longrightarrow S_2O_3^{2-}$ 9. (a) S_x^{2-} (b) $S_4O_6^{2-}$

10. When ΔG_f^o is positive, too little product is produced at equilibrium. 11. (a) N^{3-} (b) P^{3-} (c) C^{4-} (d) P^{3-} 12. (a) pyramidal (b) tetrahedral (c) angular (d) icosahedral 13. (a) $SO_2 + H_2O$ (b) $NO + H_2O$ (c) $CO_2 + H_2O$ (d) $SiO_2 + H_2O$ 14. (a) H_3PO_3 (b) $NO + HNO_3$ (c) HNO_2 (d) H_2SO_4 (e) H_2CO_3 15. One of the hydrogen atoms is bonded directly to phosphorus. 16. OCl^-, because its disproportionation reaction has the largest K 17. (a) XO_4 (b) X_2O_7 (c) XO_3 18. $HClO_4$ has only one –OH group; only one Cl–O–Cl bridge can be formed. 19. (a) planar sheet or linear double chain (b) linear chain 20. SBr_4 21. Only four bonds can be formed since the second shell can hold only an octet of electrons. 22. Silicon has low energy d orbitals that can be used in bonding, thereby making attack on Si possible; carbon does not. 23. The fluorines are so tightly packed about the S atom that they shield S from attack. 24. Branching and cross-linking (bridging of adjacent polymer strands) can occur.

20 THE TRANSITION ELEMENTS

This chapter deals with the chemistry of the elements in the center of the periodic table and those in the two rows that are placed just below the main body of the table.

20.1 GENERAL PROPERTIES

Objectives

To examine some general trends and relationships among physical and chemical properties of the transition elements.

Review

Learn the nomenclature used to refer to the different types and classes of transition elements. Review the properties that most transition metals have in common:

(a) multiple oxidation states
(b) many compounds are paramagnetic
(c) many compounds are colored
(d) strong tendency to form complex ions

New Terms

d-block elements
main transition elements
inner transition elements
triad
aqua regia

20.2 ELECTRONIC STRUCTURE AND OXIDATION STATES

Objectives

To review the electronic structures of the transition elements and to examine trends in the stabilities of oxidation states. You should learn the maximum oxidation states observed in the various B-groups and the relative stabilities of high and low oxidation states.

Review

The electronic structures of the transition elements are given in Tables 20.2 to 20.4. Note the irregularities at Cr and Cu in the first transition series. This is accounted for on the basis of the added stability possessed by a half-filled or filled subshell.

The oxidation states for the d-block elements are summarized in Tables 20.5 and 20.6. There are several points you should know.

(1) All elements in the first transition series (except Sc) show a +2 oxidation state.

(2) For Groups IIIB to VIIB the maximum positive oxidation state is equal to the group number. This also applies to the Group IIB elements and to most (but not all) of the Group IB elements.

(3) In a given period the higher oxidation states become less stable (relative to the lower ones), moving from left to right.

(4) In a given group the higher oxidation states become more stable, moving from top to bottom.

Self-Test

1. Predict the electron configurations of the following elements:

(a) Ti __________ (d) Co __________

(b) Cr __________ (e) Cu __________

(c) Mn __________

2. Why do nearly all first transition series elements show a +2 oxidation state? ____________________

3. Choose the compound in each pair that is expected to be the best oxidizing agent.

(a) V_2O_5 or Ta_2O_5 __________

(b) MnO_2 or TiO_2 ______________

(c) Sc_2O_3 or Mn_2O_3 ______________

New Terms

none

20.3 ATOMIC AND IONIC RADII

Objectives

To investigate trends in these two properties. Learn what is meant by "lanthanide contraction" and how it affects the chemistry of the elements of the third transition series.

Review

Only small horizontal changes in size occur among the transition elements because of the effectiveness of the d electrons at shielding the outer s electrons from the nucleus.

Large size increases occur from the first to the second series, but little or no changes occur from the second to the third series. This is due to the lanthanide contraction.

New Terms

lanthanide contraction

20.4 METALLURGY

Objectives

To learn how metals are extracted from their ores and purified to the point of being able to be put to practical use. You should learn the kinds of pretreatments given to ores, the way the metals are extracted from the ores, and the refining methods used to make them useful.

Review

The three steps in the commercial production of metals are: concentration, reduction and refining.

Concentration. Pretreatment procedures can be physical or chemical. Physical separations take advantage of differences in physical properties between the metal bearing component of the ore and the unwanted gangue (review flotation, amalgamation). Chemical separations rely on chemical properties to enrich the metal bearing component of the ore (review roasting, purification of Al_2O_3).

Reduction. This was discussed in Chapter 18. Review the details of the chemistry of the blast furnace.

Refining. This involves purification of the metal after reduction and the formation of alloys with desirable properties. Review the refining procedures used to produce steel from pig iron.

Self-Test

4. What property of Al_2O_3 is exploited in the purification of bauxite?

   ____

5. Write the chemical equation for the roasting of ZnS in air.

   ____

6. What is the active reducing agent in the blast furnace? ______________________

7. Write the chemical equation for the reaction between CaO and P_2O_5 during the formation of slag in the blast furnace.

   ____

8. Why did the open hearth process replace the Bessemer process for the production of steel? ______________________

9. What has enabled a return to a modified Bessemer process (the basic oxygen process)? ______________________

New Terms

refining	blast furnace	carbonyl
amalgam	slag	Bessemer converter
flotation	pig iron	open hearth furnace
gangue	cast iron	basic oxygen process
roasting		

20.5 MAGNETISM

Objectives

To account for the magnetic properties of the transition metals and their compounds. You should learn the origin of ferromagnetism and how it differs from paramagnetism.

Review

Paramagnetism arises from the presence of unpaired electrons in atoms, molecules or ions. Ferromagnetism appears to result from the alignment of many paramagnetic ions in domains in the solid state only. Alignment of domains produces a "permanent magnet". The formation of ferromagnetic domains seems to depend on interionic spacings.

New Terms

ferromagnetism
domain

20.6 COORDINATION COMPOUNDS

Objectives

To define what is meant by "coordination compound", to examine the kinds of compounds that are formed, and to review the terminology used to describe them.

Review

Complex ions formed from a metal ion and one or more ligands are called coordination compounds. Ligands are nearly always neutral molecules or negatively charged ions which have a pair of electrons that can be donated in a coordinate covalent bond. The metal and all ligands in the first coordination sphere are generally enclosed within brackets (e.g., $[Co(NH_3)_6]^{3+}$). Monodentate ligands provide one coordinating atom; bidentate ligands, two. A polydentate ligand is able to provide more than one donor atom. Important bidentate ligands are:

oxalate ion, $C_2O_4^{2-}$

ethylenediamine (en), $H_2N\text{-}CH_2\text{-}CH_2\text{-}NH_2$

Self-Test

10. What is the charge on the metal ion in each of the complex ions below?

(a) $[Mn(C_2O_4)_3]^{3-}$ __________

(b) $[FeCl_6]^{4-}$ __________

(c) $[Cr(en)_2Cl_2]^{+}$ __________

(d) $[Ni(CN)_4]^{2-}$ __________

(e) $[PtCl_6]^{2-}$ __________

New Terms

coordination compound
first coordination sphere
ligand
monodentate
bidentate
polydentate

20.7 COORDINATION NUMBER

Objectives

To examine the kinds of structures found when different numbers of atoms are coordinated to the central metal ion.

Review

Learn the definition of coordination number. Review the structures that are found for C.N. = 2, 4 and 6. Learn to draw the 2-dimensional representation of octahedral coordination described in Figure 20.8.

Self-Test

11. Practice drawing, on a separate piece of paper, the geometries observed most commonly for C.N. = 2, 4 and 6. Check yourself by referring to Figure 20.7 in the text.

New Terms

coordination number

20.8 NOMENCLATURE

Objectives

To learn how to name coordination compounds. You should be able to write the name, given the formula of a complex; you should be able to write the formula, given the name of the complex.

Review

Learn the nomenclature rules given on Pages 618 to 621 in the text. After you feel you know them, practice on the Self-Test below.

Self-Test

12. Name the following:

(a) $[Co(NH_3)_6]^{3+}$ ____________________

(b) $[CoBr_6]^{3-}$ ____________________

(c) $[Mn(en)_2Cl_2]^+$ ____________________

(d) $[Ni(H_2O)_6]^{2+}$ ____________________

(e) $[Fe(H_2O)_4(NH_3)_2]^{2+}$ ____________________

(f) $[Ag(CN)_2]^-$ ____________________

13. Write the formulas for the following:

(a) dichlorotetraamminenickel(II) ______________

(b) sodium dichlorobis(carbonato)cobaltate(III) ______________

(c) diamminesilver(I) ion ______________

(d) hexaaquochromium(III) sulfate ______________

(e) diaquobis(oxalato)chromate(III) ion ______________

(f) chlorotrithiocyanatodiammineaquomanganate(II) ion

____

New Terms

none

20.9 ISOMERISM AND COORDINATION COMPOUNDS

Objectives

To learn the meaning of the term "isomerism" and to see how it is
applied to coordination compounds. You should learn the different
types of isomerism described in this section.

Review

Compounds having the same formula but different structures are
called isomers. Review the following types of isomerism: ionization
isomerism, stereoisomerism, geometrical isomerism, optical isomerism.
Be sure you know the difference between cis- and trans- isomers. Remember that optical isomers are nonsuperimposable mirror images of each other.
Optically active compounds have the ability to rotate plane polarized light.
An equal mixture of optical isomers is said to be racemic.

Self-Test

14. Sketch the cis-trans isomers for (a) $[Co(H_2O)_4Cl_2]^+$ and
     (b) $[Co(C_2O_4)_2Cl_2]$.

15. Which of the following are expected to exhibit optical isomerism?

   (a) $[Co(NH_3)_6]^{3+}$ (d) cis-$[Co(en)_2(NH_3)_2]^{3+}$

   (b) trans-$[Co(C_2O_4)_2Cl_2]^{3-}$ (e) cis-$[Co(NH_3)_4Cl_2]^+$

   (c) $[Co(en)_3]^{3+}$

New Terms

isomer	polarized light
ionization isomerism	enantiomers
stereoisomerism	optical activity
geometrical isomerism	dextro-rotatory
optical isomerism	levo-rotatory
cis-	racemic
trans-	

20.10 BONDING IN COORDINATION COMPOUNDS: VALENCE BOND THEORY

Objectives

To account for the structure and magnetic properties of complex ions using the valence bond theory.

Review

To form the coordinate covalent bonds the metal must provide one empty hybrid orbital for each coordinating ligand atom. For C.N. = 6 this is d^2sp^3. If the d orbitals come from the 3d subshell, an inner orbital complex is formed. When 4d orbitals must be used, an outer orbital complex is formed. When the metal ion has four, five or six d electrons, it is necessary to choose between inner and outer orbital complexes because they lead to different magnetic properties. As a general rule, for d^4 or d^6 ions of the first transition series, inner orbital complexes are preferred except when the ligands are F^- or H_2O. For d^5 ions, outer orbital complexes are preferred except when the ligand is NO_2^- (nitro) or CN^-.

Tetrahedral complexes use sp^3 hybrid orbitals; square planar complexes use dsp^2 hybrids.

Self-Test

16. Give the orbital diagrams for the following octahedral complex ions.

(a) $[MnCl_6]^{2-}$

(b) $[FeCl_6]^{3-}$

(c) $[Co(CN)_6]^{3-}$

New Terms

inner orbital complex
outer orbital complex

20.11 CRYSTAL FIELD THEORY

Objectives

To describe a bonding theory that can explain the colors of complex ions as well as their magnetic properties.

Review

Crystal field theory considers the effect of the ligand ions (or dipoles) on the energies of the d orbitals of the metal ion. In an octahedral complex the ligands split the d orbitals into a low energy set of three orbitals (the t_{2g} level) and a high energy set of two orbitals (the e_g level). The energy difference between them is called Δ. When a complex absorbs light, an electron is raised in energy from the t_{2g} to the e_g level. The color of the light absorbed depends on the magnitude of Δ. The size of Δ is influenced by the nature of the ligands. Learn the spectrochemical series given on Page 634.

For metal ions with d^4, d^5, d^6 or d^7 configurations the magnetic properties are determined by the magnitude of Δ in relationship to the pairing energy P. Review the discussion of the factors that determine whether a low spin or high spin complex is formed.

Review the splitting patterns of the d orbitals for tetrahedral and square planar complexes. Remember that Δ_{tetr} is always much less than Δ_{oct}.

Self-Test

17. On a separate sheet of paper sketch the CFT splitting pattern for d orbitals in octahedral, square planar and tetrahedral complex ions. Check your answers by referring to Figures 20.26 and 20.28 in the text.

18. On a separate sheet of paper, indicate the electron population of the t_{2g} and e_g orbitals in low and high spin complexes for a d^7 metal ion.

19. Which complex in each pair below should absorb light of highest frequency (shortest wavelength)?

(a) $[CrCl_6]^{3-}$ or $[CrBr_6]^{3-}$ ________________

(b) $[CrCl_6]^{3-}$ or $[Cr(NH_3)_6]^{3+}$ ________________

(c) $[NiCl_6]^{4-}$ or $[Ni(NO_2)_6]^{4-}$ ________________

(d) $[Fe(NH_3)_6]^{3+}$ or $[Fe(CN)_6]^{3-}$ ________________

20. In a certain complex containing a metal ion with a d^6 configuration, the pairing energy is greater than Δ. Will this complex be low spin or high spin?

____

New Terms

crystal field theory
degeneracy
spectrochemical series
low spin complex
high spin complex

ANSWERS TO SELF-TEST QUESTIONS

1. see Table 20.2 in the text 2. From the loss of two 4s electrons 3. (a) V_2O_5 (b) MnO_2 (c) Mn_2O_3 4. Its amphoteric nature 5. $2ZnS + 3\,O_2 \longrightarrow 2ZnO + 2SO_2$ 6. CO 7. $3CaO + P_2O_5 \longrightarrow Ca_3(PO_4)_2$ 8. More uniform properties could be obtained. 9. Increased speed of analysis using emission spectra 10. (a) +3 (b) +2 (c) +3 (d) +2 (e) +4 11. see Figure 20.7 12. (a) hexaamminecobalt(III) ion (b) hexabromocobaltate(III) ion (c) dichlorobis(ethylenediamine)manganese (III) ion (d) hexaaquonickel(II) ion (e) diamminetetraaquoiron(II) ion (f) dicyanoargentate(I) ion 13. (a) $[Ni(NH_3)_4Cl_2]$ (b) $Na_3[Co(CO_3)_2Cl_2]$ (c) $[Ag(NH_3)_2]^+$ (d) $[Cr(H_2O)_6]_2(SO_4)_3$ (e) $[Cr(C_2O_4)_2(H_2O)_2]^-$ (f) $[Mn(NH_3)_2(SCN)_3(Cl)(H_2O)]^{2-}$
14. (a)

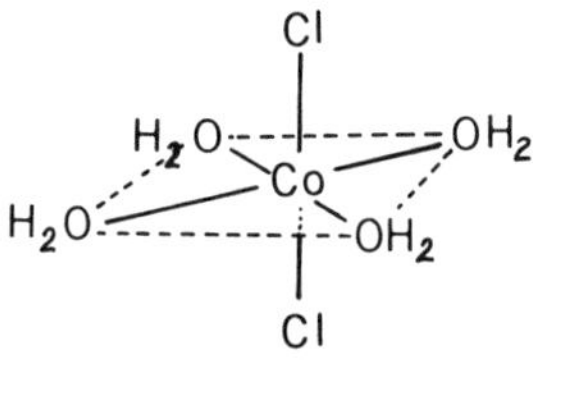

trans

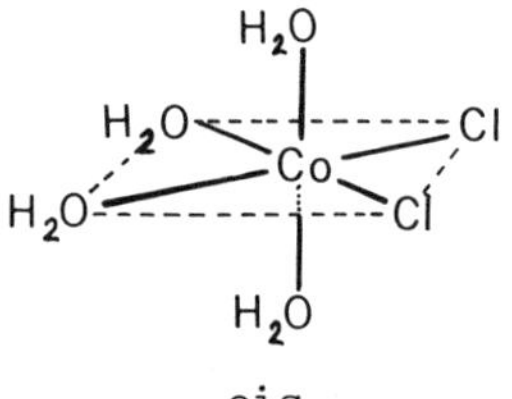

cis

(b)

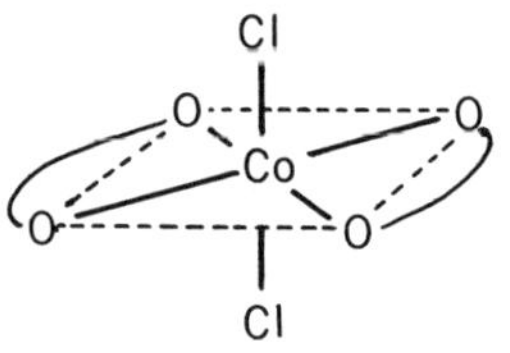

trans

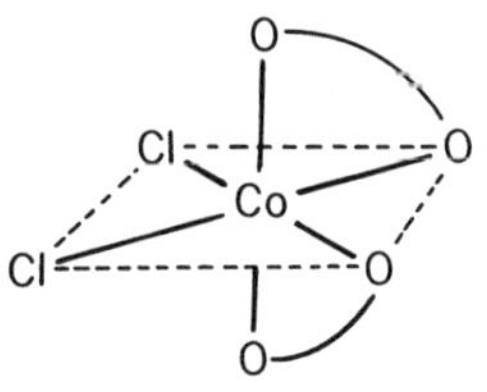

cis

15. c and d

16.

	3d					4s	4p			4d				
(a)	↑	↑	↑	xx	xx	xx	xx	xx	xx	_	_	_	_	_
(b)	↑	↑	↑	↑	↑	xx	xx	xx	xx	xx	xx	_	_	_
(c)	↑↓	↑↓	↑↓	xx	xx	xx	xx	xx	xx	_	_	_	_	_

17. see Figures 20.26 and 20.28

18.

high spin		low spin
↑ ↑	e_g	↑ _
↑↓ ↑↓ ↑	t_{2g}	↑↓ ↑↓ ↑↓

19. (a) $[CrCl_6]^{3-}$ (b) $[Cr(NH_3)_6]^{3+}$ (c) $[Ni(NO_2)_6]^{3-}$ (d) $[Fe(CN)_6]^{3-}$

20. high spin

21 ORGANIC CHEMISTRY

In this chapter we deal with the chemistry of carbon compounds. These range from simple molecules such as methane, CH_4, to extremely large polymer molecules such as polystyrene or polyethylene. The purpose of the chapter is to acquaint you with the breadth of the subject, rather than to delve too deeply into specific areas, and to illustrate some of the many practical uses to which organic compounds are applied.

21.1 HYDROCARBONS

Objectives

To examine the class of organic compounds called hydrocarbons. You should learn the nomenclature for the first ten members of the alkane, alkene and alkyne series. You should also become familiar with the bonding and structure of these compounds.

Review

Saturated hydrocarbons (the alkanes) contain only single bonds; unsaturated hydrocarbons contain either carbon-carbon double bonds (alkenes) or triple bonds (alkynes). In the alkanes the carbon is sp^3 hybridized and lies at the center of a tetrahedron. When doubly bonded, carbon employs sp^2 hybrid orbitals; when triply bonded, carbon utilizes sp hybrids.

Review the names of the alkanes in Table 21.1. You should be able to identify the number of carbon atoms in the chain by the stem of the name; for example, pentane signifies a five carbon atom chain.

Self-Test

1. Indicate the number of carbon atoms in each of the following:

(a) butene	________	(e) ethene	________
(b) hexyne	________	(f) nonene	________
(c) octane	________	(g) propyne	________
(d) methane	________	(h) heptene	________

2. Write the molecular formulas for the following:

(a) propene	________	(c) decyne	________
(b) butane	________	(d) pentane	________

3. How many π-bonds would be found in:

(a) hexene	________	(d) C_5H_8	________
(b) butylene	________	(e) C_8H_{18}	________
(c) C_6H_{12}	________		

New Terms (including the chapter introduction)

saturated
unsaturated
alkane
alkene
alkyne
homologous series
olefin
paraffin
hydrocarbon
aliphatic hydrocarbons
aromatic hydrocarbons

21.2 ISOMERS IN ORGANIC CHEMISTRY

Objectives

To examine the types of isomerism that occur among organic compounds.

Review

Structural isomers occur with the alkanes by branching of chains. Among alkenes and alkynes there are also different possible positions of the double or triple bond. Stereoisomerism includes cis-trans isomerism (geometrical isomers) and optical isomerism. Cis-trans isomers occur in alkenes where free rotation about the carbon-carbon double bond cannot occur. Optical isomers exist when a molecule contains one or more asymmetric carbon atoms (carbon atoms attached to four different groups).

Self-Test

4. On a separate sheet of paper draw the <u>cis</u> and <u>trans</u> isomers of butene. Check your answer by turning to Page 652 of the text.

5. Which of the following can exist as optical isomers?

(a)

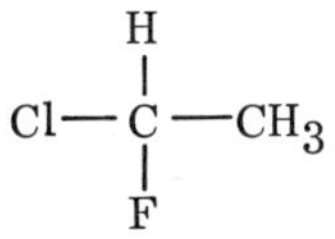

$$Cl - \underset{\underset{F}{|}}{\overset{\overset{H}{|}}{C}} - CH_3$$

(b) $CH_3 - \underset{\underset{CH_3}{|}}{CH} - CH_3$

(c) $CH_3 - CH_2 - \underset{\underset{CH_3}{|}}{CH} - CH_2 - CH_2 - CH_3$

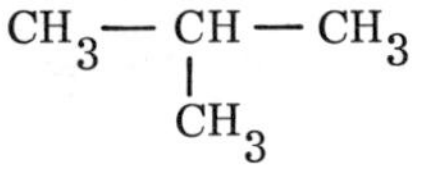

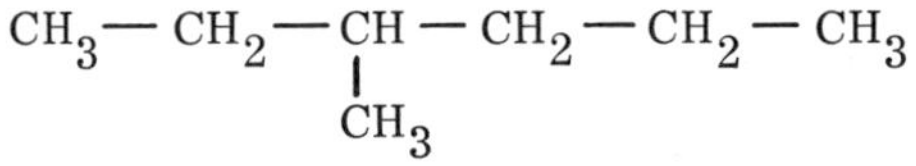

New Terms

structural isomers
asymmetric carbon atom
geometrical isomers

21.3 NOMENCLATURE

Objectives

To describe the nomenclature system that has been devised to name organic compounds. You should be able to name straight and branched-chain alkanes, alkenes and alkynes.

Review

Review the nomenclature rules and examples on Pages 655 to 658 of the text. When you feel confident that you have learned the rules, try the Self-Test for this section.

Self-Test

6. Name the compounds in Table 21.2 in your text.

7. Name the following:

(a) $CH_3 - CH = \underset{\underset{CH_3}{|}}{C} - CH_2 - CH_3$ _____

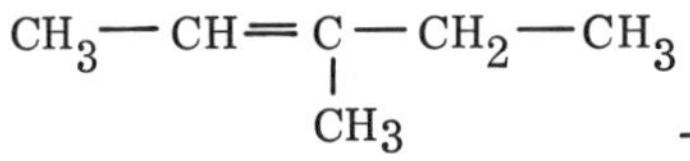

(b) $CH_3 - C = C - CH_2 - CH_3$
     / |
    CH_3 $C = CH_2$
       |
      CH_3

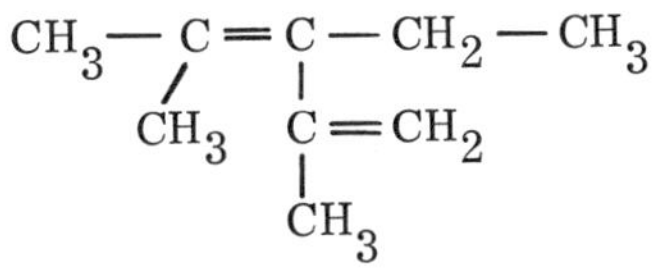

New Terms

radical

21.4 CYCLIC HYDROCARBONS

Objectives

  To examine another class of hydrocarbons.

Review

  Learn the nomenclature that is applied to these cyclic compounds. Note that their stability and structure can be related to the C–C–C angle within the ring structure.

New Terms

cyclo–

21.5 AROMATIC HYDROCARBONS

Objectives

  To study the structure, bonding and nomenclature of benzene related compounds.

Review

  Remember that benzene has a planar ring structure that can be viewed as a resonance hybrid of two structures with alternating single and double bonds. Molecular orbital theory views the bonding in terms of a delocalized π-electron cloud.

  Review the nomenclature system for benzene derivatives.

Self-Test

8. What is the C-C-C bond angle in benzene? ____________

9. Name the following:

(a)

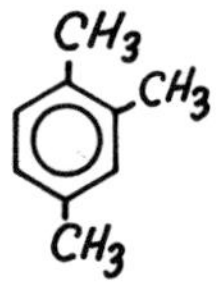

____

(b)

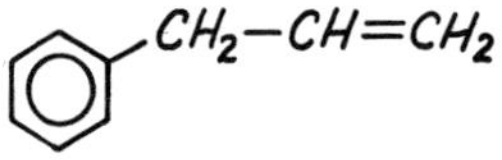

____

(c)

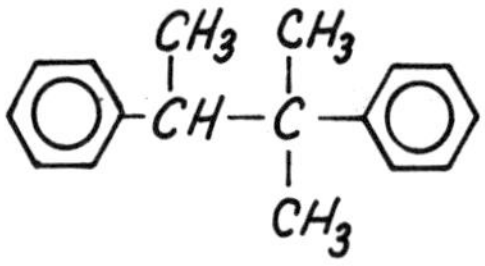

____

New Terms

ortho
meta
para
phenyl group

21.6 HYDROCARBON DERIVATIVES

Objectives

To see how most organic compounds can be considered to be derived from hydrocarbons by substituting certain groups of atoms (called functional groups) for hydrogen in the parent hydrocarbon molecule.

Review

A functional group is some group of atoms that imparts a characteristic property to an organic molecule. Review the functional groups in Table 21.5 on Page 670 of the text.

Typical reactions discussed in this section include addition reactions characteristic of unsaturated hydrocarbons. Markovnikov's rule states that when HX is added across a double bond, the H goes to the carbon atom already containing the most H's. Remember that saturated hydrocarbons primarily undergo substitution reactions.

Self-Test

10. Without referring to Table 21.5, identify the following functional groups:

(a)

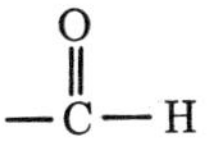

$$-\overset{\displaystyle O}{\overset{\displaystyle \|}{C}}-H$$

_____

(b) —OH

_____

(c)

$$-C\overset{\displaystyle O}{\underset{\displaystyle OH}{\diagup}}$$

_____

(d) $-\overset{|}{\underset{|}{C}}-O-\overset{|}{\underset{|}{C}}-$

_____

(e)

$$-\overset{\displaystyle O}{\overset{\displaystyle \|}{C}}-O-\overset{|}{\underset{|}{C}}-$$

_____

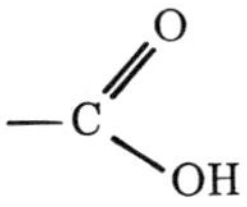

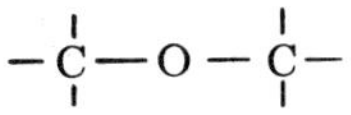

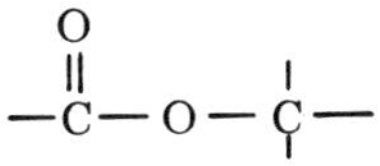

11. Write chemical equations for the following, showing the structures of the main product:

(a) $CH_3-CH=CH_2 + Br_2 \longrightarrow$ _____

(b) $CH_3-CH_3 + Cl_2 \xrightarrow{h\nu}$ _____

(c) $CH_3-\overset{\displaystyle}{\underset{\displaystyle CH_3}{C}}=CH_2 + HCN \longrightarrow$ _____

New Terms

functional group
addition reaction
substitution reaction

electrophilic addition reaction
Markovnikov's rule

21.7 HALOGEN DERIVATIVES

Objectives

To learn about some methods of preparation and characteristic reactions of halogenated hydrocarbons. You should also become aware of some of the many uses of these compounds.

Review

Many common chemicals contain halogens. Examine some of these examples on Page 672. Halogenated compounds can be prepared by substitution of alkanes and halogen addition to alkenes. The most important methods involve reactions in which the OH group of an alcohol is displaced by a halogen.

Typical reactions of halogen derivatives include nucleophilic substitution and displacement. Review these in the text.

Self-Test

12. List some uses of halogen-containing compounds. ______________

____

13. Give the products of a nucleophilic substitution reaction between the following:

(a) $CH_3{-}CH_2Cl + NH_3 \longrightarrow$ ______________

(b) $CH_3{-}CHCl{-}CH_3 + CN^- \longrightarrow$ ______________

14. Give an example of a reaction that might be used to prepare $CH_3{-}CHCl{-}CH_3$ from $CH_3{-}CHOH{-}CH_3$.

____

15. Give the products of an elimination reaction between the following:

(a) $CH_3{-}CHBr{-}CH_3 + OH^- \longrightarrow$ ______________

(b) $CH_3{-}\underset{\displaystyle CH_3}{\underset{|}{C}}Br{-}CH_3 + OH^- \longrightarrow$ ______________

New Terms

nucleophilic substitution
elimination

21.8 IMPORTANT OXYGEN-CONTAINING DERIVATIVES

Objectives

To examine some properties and chemical reactions involving a number of interrelated oxygen-containing functional groups. You should also learn some applications of the various kinds of compounds discussed in this section.

Review

Six major functional groups are discussed in this section: alcohols, aldehydes, ketones, acids, esters and ethers. Be sure you can distinguish among them.

Some important relationships in this section are:

(1) Alcohols can be oxidized to give aldehydes or ketones; aldehydes can be oxidized to give acids.

(2) The products of oxidation of an alcohol depend on whether the alcohol is primary, secondary, or tertiary.

(3) Acids react (reversibly) with alcohols to produce esters by a condensation reaction involving elimination of water. The base-catalyzed hydrolysis of an ester is called saponification.

(4) Aldehydes and ketones contain the carbonyl group, $>C{=}O$.

(5) The acidity of organic acids arises from the carboxyl group, —COOH.

(6) Condensation of two alcohols produces an ether.

Self-Test

16. Give some uses for:

(a) alcohols ____________________

(b) ketones ____________________

(c) esters ____________________

(d) ethers ____________________

17. Give the major organic product in each of the following:

(a) $$CH_3-\overset{\overset{\large O}{\|}}{C}-O-C_2H_5 + H_2O \xrightarrow{H^+}$$ ____________________

(b) $CH_3-\underset{\displaystyle OH}{\underset{|}{CH}}-CH_3 \xrightarrow{\text{oxidation}}$ ____________________

(c) $CH_3OH + CH_3COOH \longrightarrow$ ____________________

(d) $CH_3-\overset{\displaystyle O}{\overset{\|}{C}}-CH_3 + H_2 \xrightarrow{\text{catalyst}}$ ____________________

New Terms

alcohol
carbinol group
primary alcohol
secondary alcohol
tertiary alcohol
aldehyde
ketone
carbonyl group
organic acid
carboxyl group
ester
esterification
condensation reaction
saponification
ether

21.9 AMINES AND AMIDES

Objectives

To examine this class of organic compounds that can be considered to be derivatives of ammonia.

Review

Remember the functional groups for amines and amides. Learn to distinguish between primary, secondary and tertiary amines. Amines give basic aqueous solutions for the same reasons that solutions of ammonia are basic (the lone pair of electrons on the nitrogen atom of the amine is capable of accepting a proton from water).

New Terms

amine
amide
heterocyclic

21.10 POLYMERS

Objectives

To learn the general types of polymers that are formed and how their properties are related to their structure.

Review

Polymers are built by linking many small units (monomers) together to give long chains. Two polymer types exist. Addition polymers are formed by adding one monomer unit to another. Polyvinyl chloride is an example (you've probably read that vinyl chloride monomer has recently been linked to cancer). Condensation polymers are formed by elimination of a small molecule (e.g., H_2O) from between two monomer units. The polymeric oxoacids in Chapter 19 are condensation polymers.

Structural strength in polymers is improved by cross-linking between polymer chains (e.g., in Bakelite and rubber).

New Terms

monomer
addition polymer
condensation polymer
copolymer
polyester
cross-linking
vulcanization

ANSWERS TO SELF-TEST QUESTIONS

1. (a) 4 (b) 6 (c) 8 (d) 1 (e) 2 (f) 9 (g) 3 (h) 7 2. (a) C_3H_6 (b) C_4H_{10} (c) $C_{10}H_{18}$ (d) C_5H_{12} 3. (a) one (b) one (c) one (d) two (e) none 4. see Page 652 5. optical isomers found for (a) and (c) 6. (a) hexane (b) 2-methylpentane (c) 3-methylpentane (d) 2,3-dimethylbutane (e) 2,2-dimethylbutane 7. (a) 3-methyl-2-pentene (b) 2,4-dimethyl-3-ethyl-2,4-pentadiene 8. 120° 9. (a) 1,2,4-trimethylbenzene (b) 3-phenyl-1-propene (c) 2-methyl-2,3-diphenylbutane (Did you name it as a derivative of ethane?) 10. Check your answer in Table 21.5.
11. (a) $CH_3-CHBr-CH_2Br$ (b) CH_3CH_2Cl + HCl plus other substituted ethanes (c) $CH_3-C(CH_3)CN-CH_3$ 12. see Page 672
13. (a) $CH_3-CH_2NH_2 + NH_4Cl$ (HCl will react with excess NH_3) (b) $CH_3-CHCN-CH_3 + Cl^-$ 14. $CH_3-CHOH-CH_3 + HCl \longrightarrow CH_3-CHCl-CH_3 + H_2O$ 15. (a) $CH_3-CH{=}CH_2 + H_2O$

(b) $CH_3-C(CH_3)=CH_2 + H_2O$

16. (a) see Pages 674-676 (b) see Page 677 (c) see Page 680 (d) see Page 681

17. (a) $CH_3COOH + C_2H_5OH$ (b) $CH_3-CO-CH_3$ (c) $CH_3-COO-CH_3$
(d) $CH_3-CHOH-CH_3$

22 BIOCHEMISTRY

Biochemistry is the study of the chemical reactions that take place in living systems, and it is one of the most important areas of chemical research today. The remarkable breakthroughs that have occurred over the past 20 years or so have produced an impressive list of Nobel Prize winners. In this chapter we examine four important kinds of biomolecules. Read the chapter with an eye toward understanding how these substances are usually composed of simple building blocks and how the structures of biomolecules control their properties and biochemical activity.

22.1 PROTEINS

Objectives

To learn how proteins are constructed from amino acids and how protein molecules twist and bend to assume shapes that control their biological functions.

Review

The important ideas developed in this section are the following:

(1) The nature of an α-amino acid. What are the main features of an α-amino acid?

(2) The peptide bond. What is it? How is it formed?

(3) The primary structure of a protein. This is the sequence of amino acids in the peptide chain.

(4) The secondary structure of a protein. The α-helix is an example. How is the structure held in place?

(5) The tertiary structure of a protein. This is found in globular proteins and is controlled by hydrogen bonding, ionic interactions, interactions of nonpolar groups with the polar solvent (water), and the formation of disulfide bridges between cysteine molecules at different places along the chain.

(6) The quaternary structure of some proteins. This concerns the packing of globular proteins into more complex structures.

In addition, the structure of heme is described. The main thing to remember is that the heme group contains an iron atom held in a square planar ligand called a porphyrin. A similar structure is found for chlorophyll.

Self-Test (Use a separate sheet of paper to answer the following three questions.)

1. Sketch the general structure of an α-amino acid.
2. Draw the backbone for the primary structure of a tripeptide.
3. What is a disulfide bridge?

New Terms

protein
α-amino acid
peptide
peptide bond
polypeptide
primary structure
secondary structure
tertiary structure
quaternary structure
disulfide bridge
porphyrin

22.2 ENZYMES

Objectives

To see how proteins serve as catalysts to promote very specific biochemical reactions.

Review

The most important idea developed here is the "lock and key" relationship between the enzyme and the molecule upon which it acts (the substrate). This is illustrated for chymotrypsin, but holds for other enzyme-substrate interactions as well.

Review the mechanism of enzyme inhibition.

Self-Test

4. How is an enzyme irreversibly poisoned?

____

5. What is competitive inhibition?

____

New Terms

enzyme	substrate
coenzyme	inhibition

22.3 CARBOHYDRATES

Objectives

To examine the structure and properties of the class of compounds called carbohydrates.

Review

These substances are called carbohydrates because their formulas are often of the form, $C_n(H_2O)_m$. They are in fact condensation polymers of monosaccharides (polyhydroxy alochols also containing an aldehyde or ketone functional group). You should know that the monosaccharides usually exist in cyclic structures. Glucose, one of the most important monosaccharides, can exist in two forms (α-D-glucose and β-D-glucose).

Sucrose is a disaccharide. You should know the general features of the glycoside linkage.

Starch and cellulose are polymers of glucose units. Starch is formed from α-D-glucose units; cellulose from β-D-glucose units.

Self-Test

6. What are the molecular formulas for the following?

(a) sucrose _______________ (b) glucose _______________

7. What is the empirical formula for amylose (starch)? ________________

8. On a separate sheet of paper, sketch the formation of a glycoside linkage. Compare your answer to Figure 22.16 on Page 707.

9. What characterizes the names of carbohydrates?

____

New Terms

carbohydrate	polysaccharide
monosaccharide	glycoside linkage

22.4 LIPIDS

Objectives

To study the nature of this class of compounds and to see how lipids are employed by organisms.

Review

Lipids are water insoluble substances found in fats and in cell membranes. Neutral lipids are esters of glycerol and fatty acids (long hydrocarbon chains with a carboxyl group on one end). Saponification of a triglyceride gives glycerol and the anions of the fatty acids. The latter constitute a soap. Review micelle formation in solutions of soap.

Phospholipids contain two fatty acid molecules plus phosphoric acid esterified to glycerol. The phosphoric acid is also esterified to another alcohol. Phospholipids are an important component of cell membranes, forming a bilayer structure.

Another class of lipids are steroids which possess a fused ring system and display very high biological activity.

Self-Test (Use a separate sheet of paper to answer the next four questions.)

10. Draw the general structure of a nonpolar lipid. Write a reaction to show what happens when the lipid molecule is saponified. Check your answer on Pages 708 and 709 of the text.

11. Sketch a micelle formed from anions of fatty acids. You can check your answer on Page 710 of the text.

12. Sketch a bilayer structure typically formed by phospholipids. Check your answer on Page 711.

13. Draw the fused ring structure characteristic of steroids. Check your answer on Page 712.

New Terms

lipid	micelle
fatty acid	phospholipid
triglyceride	bilayer structure
soap	steroid

22.5 NUCLEIC ACIDS

Objectives

To study the composition of nucleic acids (DNA and RNA) and to see how the double helix structure of DNA can account for the transmission of genetic information from one generation to another.

Review

Key points to learn from this section are that nucleic acids are composed of three parts: a nitrogenous base, a five-carbon sugar (ribose or deoxyribose), and phosphoric acid. You should familiarize yourself with the structures of ribose and deoxyribose.

A nucleoside is formed from the nitrogenous base and the sugar. Addition of the phosphoric acid gives a nucleotide - the monomer unit in the DNA or RNA chain. Nucleotides are linked to give the nucleic acid.

In DNA two nucleic acid strands intertwine to give the double helix. These are matched and held together by base-pairing through hydrogen bond formation. Replication involves untwining the two strands and building new complementary strands against each of them, in which the original nucleic acid strands serve as templates.

Self-Test (Use a separate sheet of paper to answer the next two questions. Check your answers by referring to the text.)

14. Sketch the structure of ribose and deoxyribose.

15. Sketch the structure of (a) a nucleoside; (b) a nucleotide.

16. Which bases are able to pair in DNA? ______________________________

____

New Terms

DNA
RNA
nucleotide
nucleoside
nucleic acid
ribose
deoxyribose
double helix

22.6 PROTEIN SYNTHESIS

Objectives

To learn how DNA in the nucleus of a cell serves to provide the key to determining the primary structure of proteins.

Review

Study the functions of DNA, mRNA, and tRNA as well as the pairing of bases between DNA and RNA. Notice that uracil is found in RNA instead of thymine.

Self-Test

17. Use the DNA/mRNA base pairing scheme to deduce the base sequence that would occur in mRNA if the following base sequence occurred in DNA.

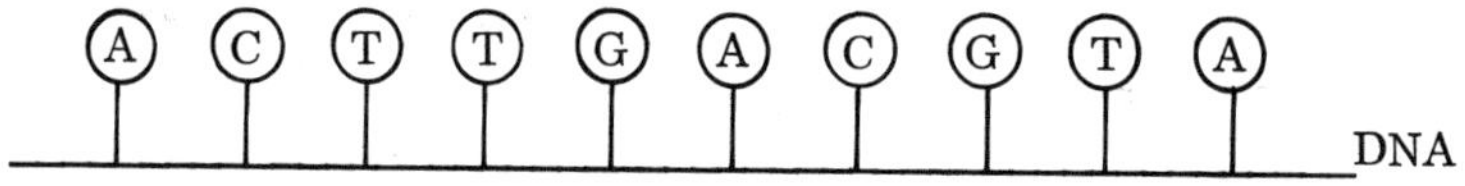

18. Use the genetic code in Table 22.3 to identify the amino acid sequence that could be constructed from mRNA with the following base sequence.

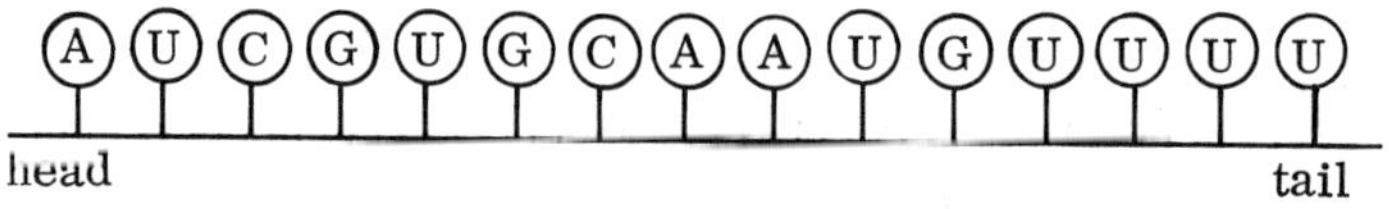

19. What amino acid sequence would occur if the fourth base were removed from the mRNA in Question 18?

____

New Terms

messenger RNA   genetic code
transfer RNA   genetic disease
codon

ANSWERS TO SELF-TEST QUESTIONS

1. $R-CH-COOH$
    $\quad\ \ |$
    $\quad\ NH_2$

2. $H_2N-CH-C-NH-CH-C-NH-CH-C-OH$
    $\qquad\ |\quad\ ||\qquad\ |\quad\ ||\qquad\ |\quad\ ||$
    $\qquad R_1\quad O\qquad R_2\quad O\qquad R_3\quad O$

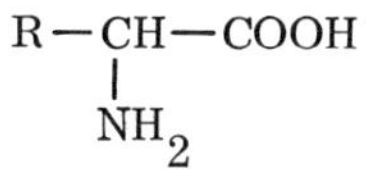

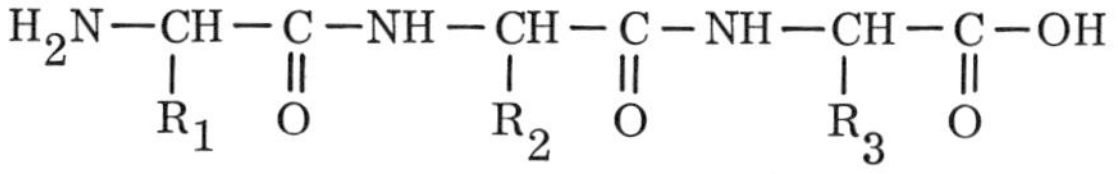

3. $R_1 - S - S - R_2$ where R_1 and R_2 belong to different parts of the protein backbone

4. When the inhibitor becomes permanently bound to the enzyme active site, the enzyme molecule becomes inoperative.   5. There is a competition between the substrate and inhibitor for the active site on the enzyme.

6. (a) $C_{12}H_{22}O_{11}$, or $C_{12}(H_2O)_{11}$ (b) $C_6H_{12}O_6$ or $C_6(H_2O)_6$

7. $C_6H_{10}O_5$ or $C_6(H_2O)_5$ 8. See Figure 22.16   9. They end in <u>ose</u> (e.g., sucr<u>ose</u>)   10. See Page 708 and 709   11. See Page 710   12. See Page 711   13. See Page 712   14. See Page 714   15. See Page 714   16. cytosine and quanine (C and G); thymine and adenine (T and A)

17.

mRNA

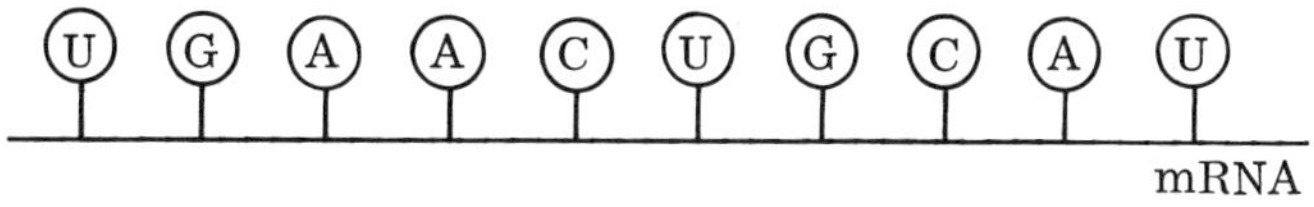

18. ile-val-gln-cys-phe   19. ile-cys-asn-val

23 NUCLEAR CHEMISTRY

In this final chapter we will examine nuclear changes, what their implications are in chemistry, and how they can be harnessed in energy production. This latter topic, of course, has become an important issue lately because of growing demands for energy and the increasing price of energy-rich fuels.

23.1 SPONTANEOUS RADIOACTIVE DECAY

Objectives

To review the types of nuclear decay processes that occur and to learn how to write nuclear equations.

Review

You should be sure you know the mass, charge and symbol for each of the particles in Table 23.1. If necessary, review the method of indicating mass number and atomic number for isotopes in Section 3.9 (Page 57 of the text, Page 43 of this Study Guide).

In balancing a nuclear equation, remember that mass and charge must both be balanced. This requires that the sum of mass numbers (left superscript) on both sides of the arrow must be equal. Also, the algebraic sum of charges (left subscripts) on both sides of the arrow must be the same. Review Example 23.1 in the text.

Kinetically, radioactive decay is a first order process. The relative rate of decay of radioisotopes are usually expressed in terms of the half-life, $t_{\frac{1}{2}}$, which is the time required for half of the sample to decay. Learn to apply Equations 23.1 and 23.2 (review Examples 23.2 and 23.3).

Archaeological dating is described as an application of radioactive decay process. Review Example 23.4.

Self-Test

1. Without referring to the text, write the symbols for the following. Afterward, check your answer by referring to Table 23.1 in the text.

   (a) proton ________ (d) alpha particle ________

   (b) neutron ________ (e) positron ________

   (c) beta particle ________

2. Fill in the missing symbol in each of the following:

   (a) ${}^{27}_{13}Al + {}^{4}_{2}He \longrightarrow$ ________ $+ {}^{1}_{0}n$

   (b) ${}^{63}_{29}Cu + {}^{1}_{1}H \longrightarrow$ ________

   (c) ${}^{133}_{56}Ba + {}^{0}_{-1}e \longrightarrow$ ________

   (d) ${}^{213}_{83}Bi \longrightarrow {}^{209}_{81}Tl +$ ________

   (e) ${}^{209}_{82}Pb \longrightarrow {}^{209}_{83}Bi +$ ________

   (f) ${}^{238}_{92}U +$ ________ $\longrightarrow {}^{247}_{99}Es + 5\,{}^{1}_{0}n$

   (g) ${}^{235}_{92}U + {}^{1}_{0}n \longrightarrow {}^{139}_{36}Ba +$ ________ $+ 3\,{}^{1}_{0}n$

3. The half-life of ${}^{90}_{38}Sr$ is 19.9 years. What is the rate constant for the decay process? ________

4. A piece of wood taken from a burial mound of ancient civilization was found to contain only 1/3 as much carbon-14 as in a live tree. Estimate the age of the wood (and the burial mound). ________

New Terms

nuclide
parent and daughter isotopes
radioactive series
decay series
Geiger-Muller counter
half-life

23.2 NUCLEAR TRANSFORMATIONS

Objectives

To briefly examine how nuclear reactions can be brought about by bombarding nuclei with various particles.

Review

Nuclear transformations occur when target nuclei are bombarded with various particles. Learn the shorthand notation used to describe these reactions (e.g., $^{27}_{13}Al\ (\alpha, n)^{30}_{15}P$).

New Terms

nuclear transformation
particle accelerator
cyclotron

23.3 NUCLEAR STABILITY

Objectives

To look at what characteristics stable and unstable nuclei exhibit. You should learn the kinds of reactions unstable nuclei undergo to become stable.

Review

Stable nuclei lie in a "band of stability". With the exception of hydrogen, stable nuclei always possess at least as many neutrons as protons; at high atomic number neutrons outnumber protons. The band of stability ends at Z = 83.

Elements having an n/p ratio that is too high generally decay by beta emission or neutron emission. Elements with low n/p ratios tend to emit positrons or undergo electron capture. Elements with Z > 83 undergo alpha emission or nuclear fission.

Stable nuclei with odd numbers of both protons and neutrons are rare. Conversely, stable nuclei with even numbers of protons and neutrons are much more common. Some nuclei are very stable and possess "magic numbers" of both protons and neutrons. This supports the nuclear shell theory.

Self-Test

5. The isotope $^{107}_{48}Cd$ decays by electron capture. Write the nuclear equation for the reaction.

____

6. The isotope $^{107}_{49}In$ lies below the band of stability. Which of the following is a likely decay mode for this isotope?

(a) $^{107}_{49}In \longrightarrow \, ^{107}_{50}Sn + \, ^{0}_{-1}e$ (c) $^{107}_{49}In \longrightarrow \, ^{103}_{47}Ag + \, ^{4}_{2}He$

(b) $^{107}_{49}In \longrightarrow \, ^{106}_{49}In + \, ^{1}_{0}n$ (d) $^{107}_{49}In \longrightarrow \, ^{107}_{48}Cd + \, ^{0}_{1}e$

____

New Terms

band of stability
electron capture
K capture
fission
magic number

23.4 EXTENSION OF THE PERIODIC TABLE

Objectives

To consider the possibility of stable superheavy elements beyond the current range of the periodic table.

Review

The possibility of very heavy stable nuclei is predicted on the basis of magic numbers of 114 for protons and 184 for neutrons.

Self-Test

7. How many electrons could be accommodated in a g subshell? __________

8. Which element (currently known) would be expected to have chemical properties most similar to element Z = 114?

____

9. If element 115 were isolated, what would you expect the formula of its oxide to be?

____

New Terms

none

23.5 CHEMICAL APPLICATIONS

Objectives

To examine some chemical applications of radioactive isotopes.

Review

Chemical applications of radioisotopes rely primarily on their ability to be detected and counted. This section illustrates several examples of applications to chemical analysis, the study of descriptive chemistry, and the study of reaction mechanisms.

Review the reasoning involved in the isotope dilution method. Apply this reasoning to Question 10 in the Self-Test. Also review the principles of neutron activation analysis.

Self-Test

10. A 1.00-g portion of KCl labeled with ^{40}K and having a specific gravity of 100 cpm per gram was added to a 112-g sample of a mixture of KCl and other salts. By fractional crystallization some pure KCl was recovered from the mixture. This KCl had a specific activity of 1.2 cpm per gram. What weight of KCl was in the original 112-g sample? What percent of the sample was KCl?

11. A student proposed the following structure for the thiosulfate, $S_2O_3^{2-}$, ion: $[O-S-O-S-O]^{2-}$ Thiosulfate is formed from SO_3^{2-} by reaction with elemental sulfur.

$$S + SO_3^{2-} \longrightarrow S_2O_3^{2-}$$

If radioactive sulfur is used in this reaction, the $S_2O_3^{2-}$ becomes labeled. When treated with H^+, the $S_2O_3^{2-}$ decomposes and SO_2 is evolved. None of the radioactivity occurs in this SO_2. How does this argue against the structure proposed by the student?

New Terms

tracer study
isotope dilution
neutron activation analysis

23.6 NUCLEAR FISSION AND FUSION

Objectives

To examine nuclear fission and fusion processes and to consider them as potential sources of energy.

Review

Fission chain reactions occur because more neutrons are produced during fission than are consumed. The minimum amount of fissionable material required to sustain a chain reaction is called the critical mass.

Review the operation of a nuclear reactor. A breeder reactor produces more fuel than it consumes.

Nuclear fusion involves the creation of a heavier nucleus from two lighter ones. Fusion liberates a large amount of energy but requires extremely high temperatures to occur.

New Terms

fission
fusion
critical mass
breeder reactor
plasma

23.7 NUCLEAR BINDING ENERGY

Objectives

To investigate the source of the large energy changes that occur during fission and fusion.

Review

The mass of a given nucleus is always less than the sum of the masses of the individual protons and neutrons that go toward forming the nucleus. The difference between the calculated and actual masses is called the mass defect. Its energy equivalent is called the binding energy and can be calculated from the mass defect by applying Einstein's equation, $E = mc^2$. In these calculations, illustrated on Page 745, use the value,

$$c = 2.9979 \times 10^{10} \text{ cm/sec}$$

The binding energy can be expressed in MeV, in which case,

1 amu = 931 MeV

The highest binding energy per nucleon occurs in the vicinity of iron in the periodic table. Fission releases energy because the lighter particles produced have greater binding energy. Fusion releases energy because the heavier particles formed have much larger binding energies.

Self-Test

12. (a) Compute the binding energy for $^{32}_{16}S$ which has an atomic mass of 31.97207 amu. __________

(b) What is the binding energy per nucleon for $^{32}_{16}S$? __________

New Terms

mass defect
binding energy
MeV

ANSWERS TO SELF-TEST QUESTIONS

1. See Table 23.1 2. (a) $^{30}_{15}P$ (b) $^{64}_{30}Zn$ (c) $^{133}_{55}Cs$ (d) $^{4}_{2}He$ (e) $^{0}_{-1}e$ (f) $^{14}_{7}N$ (g) $^{139}_{36}Kr$ 3. $k = 3.48 \times 10^{-2}\ yr^{-1}$ 4. 9000 years 5. $^{107}_{48}Cd + ^{0}_{-1}e \longrightarrow ^{107}_{47}Ag$ 6. d 7. 18 8. Pb 9. X_2O_3 10. 83.3 g KCl, 74.4% 11. Since either bond 1 or 2 in O — S $\overset{1}{—}$ O $\overset{2}{—}$ S — O would be expected to break with equal ease, there is no reason to expect that this structure would <u>never</u> give the labeled sulfur in the SO_2. An unsymmetrical structure is required to explain the results of this experiment. The actual structure of $S_2O_3^{2-}$ is

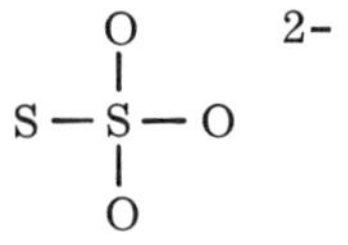

12. (a) 272 MeV (b) 8.50 MeV/nucleon

DETAILED SOLUTIONS TO SELECTED EVEN-NUMBERED PROBLEMS FROM THE TEXTBOOK

On the pages that follow you will find worked-out, detailed solutions to a representative sampling of the even-numbered problems that appear at the end of the chapters in the textbook. The purpose of these is to help you learn how to solve problems. For them to be useful to you, however, you must devote your own effort at solving them. Therefore, before referring to a particular solution, attempt to work the problem by yourself. If you can't solve the problem, then refer to these solutions. After you understand how a particular problem is solved, try another one similar to it to be sure you've mastered the material.

Theodore W. Sottery

1.20 (a) 1.25×10^3 (b) 1.3×10^7 (c) 6.023×10^{22} (d) 2.1457×10^5

1.24 (a) 5.56×10^3 (b) 2.9×10^4 (c) 1.49×10^{10} (d) 3.8×10^{-6}

1.28 (a) $? \text{ cm} = 1.40 \text{ m} \times \dfrac{10^2 \text{ cm}}{1 \text{ m}} = \underline{\underline{140 \text{ cm}}}$

(c) $? \ell = 185 \text{ ml} \times \dfrac{1 \ell}{10^3 \text{ ml}} = \underline{\underline{0.185 \ell}}$ (e) $? \text{m}^2 = 10 \text{ yd}^2 \times \left(\dfrac{3 \text{ ft}}{1 \text{ yd}}\right)^2 \times \left(\dfrac{1 \text{ m}}{3.281 \text{ ft}}\right)^2 = \underline{\underline{8 \text{ m}^2}}$

(g) $\dfrac{? \text{ mi}}{1 \text{ hr}} = \dfrac{20 \text{ ft}}{1 \text{ sec}} \times \dfrac{1 \text{ mi}}{5280 \text{ ft}} \times \dfrac{60 \text{ sec}}{1 \text{ min}} \times \dfrac{60 \text{ min}}{1 \text{ hr}} = \underline{\underline{14 \text{ mi/hr}}}$

(i) $\dfrac{? \text{ cm}}{1 \text{ sec}} = \dfrac{40 \text{ mi}}{1 \text{ hr}} \times \dfrac{1 \text{ hr}}{60 \text{ min}} \times \dfrac{1 \text{ min}}{60 \text{ sec}} \times \dfrac{1609 \text{ m}}{1 \text{ mi}} \times \dfrac{10^2 \text{ cm}}{1 \text{ m}} = \underline{\underline{\dfrac{2 \times 10^3 \text{ cm}}{1 \text{ sec}}}}$

1.30 $\dfrac{? \text{ km}}{1 \text{ hr}} = \dfrac{35 \text{ mi}}{1 \text{ hr}} \times \dfrac{1.609 \text{ km}}{1 \text{ mi}} = \underline{\underline{56 \text{ km/1 hr}}}$

1.34 (I) $10 \text{ g Cu} \leftrightarrow 1.26 \text{ g O}$ (II) $10 \text{ g Cu} \leftrightarrow 2.52 \text{ g O}$

$$\frac{1.26 \text{ g O (I)}}{2.52 \text{ g O (II)}} = \frac{1 \text{ (I)}}{2 \text{ (II)}}$$

1.36 K.E. = $1/2\ mv^2$

$? \text{ J} = \dfrac{1}{2} \times 4500 \text{ kg} \left(\dfrac{1.79 \text{ m}}{1 \text{ sec}}\right)^2 = \underline{7.2 \times 10^3 \text{ kg m}^2 \text{sec}^{-2}} = \underline{\underline{7.2 \times 10^3 \text{ J}}}$

$? \text{ cal} = 7.2 \times 10^3 \text{ J} \times \dfrac{1 \text{ cal}}{4.184 \text{ J}} = \underline{\underline{1.7 \times 10^3 \text{ cal}}}$

1.38 $? °F = 30 °C \times \frac{9 F°}{5 C°} + 32 F° = \underline{\underline{86 °F}}$

$? F° = 1983 °C \times \frac{9 F°}{5 C°} + 32 F° = \underline{\underline{3601 °F}}$

1.44 $? °C = (6152 °F - 32 F°) \times \frac{5 C°}{9 F°} = \underline{\underline{3400. °C}}$

2.12 (a) 40.31 (b) 110.98 (c) 208.22 (d) 135.02 (e) 163.94

2.14 $?\,g(C) = 1.35\ mol(C) \times 194\ g(C)/1\ mol(C) = 262\ g(C)$

2.18 $?\,mol\ NaHCO_3 = 242\ \cancel{g\ NaHCO_3} \times \frac{1\ mol\ NaHCO_3}{84.0\ \cancel{g\ NaHCO_3}} =$
$2.88\ mol\ NaHCO_3$

2.24 $?\,mol\ BaSO_4 = 1.25\ \cancel{mol\ Al_2(SO_4)_3} \times \frac{3\ mol\ BaSO_4}{1\ \cancel{mol\ Al_2(SO_4)_3}} = 3.75\ mol\ BaSO_4$

2.26 $?\,mol\ S = 632\ \cancel{g\ FeS_2} \times \frac{1\ \cancel{mol\ FeS_2}}{119.97\ \cancel{g\ FeS_2}} \times \frac{2\ mol\ S}{1\ \cancel{mol\ FeS_2}} = 10.5\ mol\ S$

2.30 $?\,g\,C = 3\ \cancel{cm\ C} \times \frac{1\ \cancel{at.\ C}}{1.5 \times 10^{-8}\ \cancel{cm\ C}} \times \frac{12.01\ g\ C}{6.022 \times 10^{23}\ \cancel{at\ C}} = 3.99 \times 10^{-15}\ g\ C$

2.32 $\%X = gX/g\,Cpd \times 100$ (a) 34.43 % Fe, 65.57 % Cl
(b) 42.07 % Na, 18.89 % P, 39.04 % O (c) 28.71 % K, 0.74 % H,
23.55 % S, 47.00 % O (d) 21.21 % N, 6.87 % H, 23.45 % P, 48.46 % O
(e) 84.98 % Hg, 15.02 % Cl

2.34 $?\,g\,N = 30\ \cancel{g(Gl)} \times \frac{14\ g\ N}{75\ \cancel{g(Gl)}} = 5.6\ g\ N$

2.39 $C_xH_yO_z$ $? \text{ mol C} = 63.2 \text{ gC} \times \frac{1 \text{ mol C}}{12.0 \text{ gC}} = \underline{5.27 \text{ mol C}}$

$? \text{ mol H} = 5.26 \text{ gH} \times \frac{1 \text{ mol H}}{1.01 \text{ gH}} = \underline{5.22 \text{ mol H}}$ $? \text{ mol O} = 31.6 \text{ gO} \times \frac{1 \text{ mol O}}{16.0 \text{ gO}} = \underline{1.98 \text{ m.O}}$

$\div 1.98$ $\therefore \underline{2.67} : \underline{2.64} : \underline{1}$ OR $\times 3$ $\underline{8.01} : \underline{7.92} : \underline{3}$ $\therefore C_8H_8O_3$

2.40 $? \text{ mol C} = 1.030 \text{ g}CO_2 \times \frac{1 \text{ mol } CO_2}{44.01 \text{ g}CO_2} \times \frac{1 \text{ mol C}}{1 \text{ mol } CO_2} = \underline{0.02341 \text{ mol C}}$

$? \text{gC} = 0.02341 \text{ mol C} \times 12.01 \text{ gC}/1 \text{ mol C} = \underline{0.2811 \text{ gC}}$

$? \text{ mol H} = 0.632 \text{ g}H_2O \times \frac{1 \text{ mol } H_2O}{18.02 \text{ g}H_2O} \times \frac{2 \text{ mol H}}{1 \text{ mol } H_2O} = \underline{0.07016 \text{ mol H}} \times \frac{1.008 \text{ gH}}{1 \text{ mol H}} =$
$\underline{0.07072 \text{ gH}}$

$? \text{gO} = 0.537 \text{ g Cpd} - 0.2811 \text{ gC} - 0.07072 \text{ gH} = \underline{0.185 \text{ gO}} \times \frac{1 \text{ mol O}}{16.00 \text{ gO}} =$
$\underline{0.0116 \text{ mol O}}$ ($\div$ mol of C, H & O by 0.0116)

$\therefore \underline{2.02 (C)} : \underline{6.06 (H)} : \underline{1.00 (O)}$ $\therefore C_2H_6O$

2.46 $? \text{g(SA)} = 2 \times 5 \text{ grain(A)} \times \frac{1 \text{g}}{15.4 \text{ grains}} \times \frac{1 \text{ mol(A)}}{180.0 \text{ g(A)}} \times \frac{1 \text{ mol(SA)}}{1 \text{ mol(A)}} \times \frac{138 \text{ g(SA)}}{1 \text{ mol(SA)}}$
$= \underline{0.50 \text{ g(SA)}}$

2.48 (a) $? \text{ mol } HClO_3 = 14.3 \text{ g}ClO_2 \times \frac{1 \text{ mol } ClO_2}{67.46 \text{ g}ClO_2} \times \frac{5 \text{ mol } HClO_3}{6 \text{ mol } ClO_2} =$
$\underline{0.177 \text{ mol } HClO_3}$

(b) $? \text{g}H_2O = 5.74 \text{ gHCl} \times \frac{1 \text{ mol HCl}}{36.46 \text{ gHCl}} \times \frac{3 \text{ mol } H_2O}{1 \text{ mol HCl}} \times \frac{18.00 \text{ g}H_2O}{1 \text{ mol } H_2O} = \underline{8.50}$ $(\text{g}H_2O)$

(c) $? \text{g}HClO_3 = 4.25 \text{ g}ClO_2 \times \frac{5 \times 84.46 \text{ g}HClO_3}{6 \times 67.46 \text{ g}ClO_2} = \underline{4.43 \text{ g}HClO_3}$ ← (ClO_2 is the Limiting reagent)

$? \text{g}HClO_3 = 0.853 \text{ g}H_2O \times 5 \times 84.46 \text{ g}HClO_3 / 3 \times 18.00 \text{ g}H_2O = \underline{6.67 \text{ g}HClO_3}$

2.52

(a) $? \text{ mol HCl} = 0.430 \text{ mol } COCl_2 \times \frac{2 \text{ mol HCl}}{1 \text{ mol } COCl_2} = \underline{\underline{0.860 \text{ mol HCl}}}$

(b) $? \text{ g HCl} = 11.0 \text{ g } CO_2 \times \frac{1 \text{ mol } CO_2}{44.0 \text{ g } CO_2} \times \frac{2 \text{ mol HCl}}{1 \text{ mol } CO_2} \times \frac{36.46 \text{ g HCl}}{1 \text{ mol HCl}} = \underline{\underline{18.2 \text{ g HCl}}}$

(c) $? \text{ mol HCl} = 0.200 \text{ mol } COCl_2 \times 2 \text{ mol HCl} / 1 \text{ mol } COCl_2 = \underline{\underline{0.400 \text{ mol HCl}}}$

(Limiting reagent)

$? \text{ mol HCl} = 0.400 \text{ mol } H_2O \times 2 \text{ mol HCl} / 1 \text{ mol } H_2O = \underline{0.800 \text{ mol HCl}}$

2.54

(a) $? \text{ g HCl} = 35.0 \text{ g } C_2H_2 \times \frac{1 \text{ mol } C_2H_2}{26.0 \text{ g } C_2H_2} \times \frac{1 \text{ mol HCl}}{1 \text{ mol } C_2H_2} \times \frac{36.5 \text{ g HCl}}{1 \text{ mol HCl}} =$ 49.1 g HCl

→ Since 51.0 g of HCl are available C_2H_2 is Limiting

(b) $? \text{ g } C_2H_3Cl = 35.0 \text{ g } C_2H_2 \times 62.46 \text{ g } C_2H_3Cl / 26.00 \text{ g } C_2H_2 = \underline{\underline{84.1 \text{ g } C_2H_3Cl}}$

(c) XS HCl = 51.0 g (supplied) − 49.1 g (consumed) = **1.90 g HCl (XS)**

2.62

$? \text{ ton } H_2SO_4 = 25.0 \text{ ton } Ca_3(PO_4)_2 \times \frac{2 \times 98.00 \text{ ton } H_2SO_4}{1 \times 310.19 \text{ ton } Ca_3(PO_4)_2} = \underline{\underline{15.8 \text{ ton } H_2SO_4}}$

3.56 $\dfrac{?\ g\ Eu\ (AV)}{1\ mol\ Eu} = \dfrac{0.4782 \times 150.9\ g\ Eu(151) + 0.5218\ g\ Eu(153)}{1\ mol\ Eu} = \underline{\underline{151.9\ \dfrac{g\ Eu}{mol\ Eu}}}$

insert $\times 152.9$

3.58 $\dfrac{?\ g\ Pb(AV)}{1\ mol\ Pb} = (204.0 \times 0.0148 + 206.0 \times 0.236 + 207.0 \times 0.226 + 208.0 \times 0.523)\ g\ Pb(AV)/1\ mol\ Pb = \underline{\underline{207\ \dfrac{g\ Pb(AV)}{1\ mol\ Pb}}}$

3.62 $\dfrac{1}{\lambda} = 109{,}678\ cm^{-1}\left(\dfrac{1}{n_2^2} - \dfrac{1}{n_4^2}\right) = 109{,}678\ cm^{-1}\left(\dfrac{1}{2^2} - \dfrac{1}{4^2}\right) = \underline{20{,}564.6\ cm^{-1}}$

$\therefore\ \lambda = \dfrac{1}{20{,}564.6\ cm^{-1}} \times 10^7\ nm/1\ cm = \underline{\underline{486.272\ nm}}$ (b) $\underline{\underline{1094.11\ nm}}$

3.64 $c = \nu\lambda\ \therefore\ \lambda = c/\nu$ $\quad ?\ cm = \dfrac{2.998 \times 10^{10}\ cm\ sec^{-1}}{8.0 \times 10^{15}\ Hz} \times \dfrac{1\ Hz}{1\ sec^{-1}} \times \dfrac{10^7\ nm}{1\ cm} = \underline{\underline{37.0\ nm}}$

$\nu = c/\lambda$ $\quad ?\ Hz = \dfrac{2.998 \times 10^{10}\ cm\ sec^{-1}}{200\ nm} \times \dfrac{10^7\ nm}{1\ cm} \times \dfrac{1\ Hz}{1\ sec^{-1}} = \underline{\underline{1.5 \times 10^{15}\ Hz}}$

3.66 $E = h\nu$ $\quad ?\ erg = 6.626 \times 10^{-27}\ erg\ sec \times 3 \times 10^{15}\ Hz \times \dfrac{1\ sec^{-1}}{1\ Hz} = \underline{\underline{2 \times 10^{-11}\ erg}}$

4.52

$$K_{(s)} + \frac{1}{2}Cl_{2\,(g)} \longrightarrow KCl_{(s)}$$

$K_{(s)} \longrightarrow K_{(g)}$		21.5 kcal
$\frac{1}{2}Cl_{2\,(g)} \longrightarrow Cl_{(g)}$		28.5 kcal
$K_{(g)} \longrightarrow K^{+}_{(g)}$		100.1 kcal
$Cl_{(g)} \longrightarrow Cl^{-}_{(g)}$		−83.3 kcal
$K^{+}_{(g)} + Cl^{-}_{(g)} \longrightarrow KCl_{(s)}$		−168.3 kcal
$K_{(s)} + \frac{1}{2}Cl_{2\,(g)} \longrightarrow KCl_{(s)}$		−101.5 kcal

4.54 $E_N \sim (|I.E.| + |E.A.|)/2$ $E_N(F) \sim 241$, $E_N(Cl) \sim 192$, $E_N(Br) \sim 176$, $E_N(I) \sim 156$ $\therefore$ E_N decreases from $F \rightarrow I$

5.42 (a) $? \text{ mol KCl} = 250 \text{ ml sol} \cdot \dfrac{0.10 \text{ mol HCl}}{1000 \text{ ml sol}} = \underline{\underline{2.5 \times 10^{-2} \text{ mol KCl}}}$

(b) $? \text{ mol } HClO_4 = 1.65 \text{ l sol} \times \dfrac{1.40 \text{ mol } HClO_4}{1 \text{ l sol.}} = \underline{\underline{2.31 \text{ mol } HClO_4}}$

5.44 $? \text{ g } Ba(OH)_2 = 250 \text{ ml sol} \times \dfrac{0.300 \text{ mol } OH^-}{10^3 \text{ ml sol}} \times \dfrac{1 \text{ mol } Ba(OH)_2}{2 \text{ mol } OH^-} \times \dfrac{171 \text{ g } Ba(OH)_2}{1 \text{ mol } Ba(OH)_2}$
$= \underline{\underline{6.41 \text{ g } Ba(OH)_2}}$

5.46 $\dfrac{? \text{ g sol.}}{1 \text{ ml sol.}} = \dfrac{273.8 \text{ g } MgSO_4}{1 \text{ l. sol.}} \times \dfrac{1 \text{ l. sol.}}{10^3 \text{ ml sol.}} \times \dfrac{100 \text{ g sol}}{22.0 \text{ g } MgSO_4} = \underline{\underline{\dfrac{1.24 \text{ g sol.}}{1 \text{ ml sol.}}}}$

$\dfrac{? \text{ mol } MgSO_4}{1 \text{ l sol.}} = \dfrac{273.8 \text{ g } MgSO_4}{1 \text{ l sol.}} \times \dfrac{1 \text{ mol } MgSO_4}{120.37 \text{ g } MgSO_4} = \underline{\underline{2.275 \text{ M}}}$

5.48 $? \text{ g AgCl} = 25.0 \text{ ml sol.} \times \dfrac{0.050 \text{ mol HCl}}{10^3 \text{ ml sol.}} \times \dfrac{1 \text{ mol AgCl}}{1 \text{ mol HCl}} \times \dfrac{143.4 \text{ g AgCl}}{1 \text{ mol AgCl}} = \underline{\underline{0.179 \text{ g}}}$

$? \text{ g AgCl} = 100 \text{ ml sol.} \times \dfrac{0.050 \text{ mol } AgNO_3}{10^3 \text{ ml sol.}} \times \dfrac{1 \text{ mol AgCl}}{1 \text{ mol } AgNO_3} \times \dfrac{143.4 \text{ g AgCl}}{1 \text{ mol AgCl}} = \underline{\underline{0.717 \text{ g}}}$

$\therefore$ $\underline{\underline{\text{HCl lim. rgnt.}}}$ $\therefore$ $\underline{\underline{0.179 \text{ g AgCl}}}$

5.50 $? \text{ ml sol} = 25.0 \text{ ml (II)} \times \dfrac{0.200 \text{ mol } Fe_2(SO_4)_3}{10^3 \text{ ml (II)}} \times \dfrac{3 \text{ mol } BaCl_2}{1 \text{ mol } Fe_2(SO_4)_3} \times \dfrac{10^3 \text{ ml sol.}}{10^{-1} \text{ mol } BaCl_2} =$
$\underline{\underline{150. \text{ ml sol.}}}$

5.52 (a) $\dfrac{? \text{ g (C.A.)}}{1 \text{ mol (C.A.)}} = \dfrac{0.100 \text{ g (C.A.)}}{17.2 \text{ ml sol}} \times \dfrac{10^3 \text{ ml sol}}{0.0500 \text{ mol NaOH}} \times \dfrac{1 \text{ mol NaOH}}{1 \text{ mol (C.A.)}} = \underline{\underline{116}}$

$C_3H_6O = 58$ $\therefore$ (b) $\underline{\underline{C_6H_{12}O_2}}$

5.56 $? \text{g Cl} = 0.249\ \text{g Cpd} \times \frac{0.694\ \text{g AgCl}}{0.249\ \text{g Cpd}} \times \frac{35.45\ \text{g Cl}}{143.4\ \text{g AgCl}} = \underline{0.172\ \text{g Cl}}$

$? \text{g Ti} = 0.249\ \text{g cpd} \times \frac{(0.249 - 0.172)\ \text{g Ti}}{0.249\ \text{g cpd}} = \underline{0.077\ \text{g Ti}}$

$? \text{mol Cl} = 0.172\ \text{g Cl} \times \frac{1\ \text{mol Cl}}{35.45\ \text{g Cl}} = \underline{4.84 \times 10^{-3}}$ $? \text{mol Ti} = 0.077\ \text{g Ti} \times \frac{1\ \text{mol Ti}}{47.9\ \text{g Ti}}$
$= \underline{1.6 \times 10^{-3}\ \text{mol Ti}}$

$\therefore \underline{3\ \text{mol Cl}} : \underline{1\ \text{mol Ti}} \quad \therefore \underline{\underline{TiCl_3}}$

5.60 (a) $Mn^{2+} \longrightarrow Mn(3+) + 1e$

$? \text{g } MnSO_4 = 1\ \text{geq } MnSO_4 \times 151\ \text{g } MnSO_4 / 1\ \text{geq. } MnSO_4 = \underline{\underline{151\ \text{g } MnSO_4}}$

(c) $Mn^{2+} \longrightarrow Mn(6+) + 4e$

(b)(75.5) (d) (30.2)

$\frac{? \text{g } MnSO_4}{1\ \text{geq. } MnSO_4} = \frac{151\ \text{g } MnSO_4}{4\ \text{geq } MnSO_4} = \underline{\underline{37.8\ \text{g } MnSO_4 / 1\ \text{geq } MnSO_4}}$

5.64 $\frac{? \text{g (A)}}{1\ \text{geq.(A)}} = \frac{4.93\ \text{g(A)}}{129\ \text{ml } Ba(OH)_2} \times \frac{10^3\ \text{ml } Ba(OH)_2}{0.850\ \text{geq.(B)}} \times \frac{1\ \text{geq.(B)}}{1\ \text{geq.(A)}} = \underline{\underline{45.0\ \frac{\text{g(A)}}{\text{geq.(A)}}}}$

5.76 $? \text{ml(F)} = 100\ \text{ml(I)} \times \frac{0.500\ \text{mol}}{10^3\ \text{ml(I)}} \times \frac{10^3\ \text{ml(II)}}{0.200\ \text{mol}} = \underline{\underline{250\ \text{ml(F)}}}$

6.20

$$? \text{torr (B)} = 836 \text{ torr(A)} + 74 \text{ cm(o)} \times \frac{0.847 \text{ cm}H_2O}{1 \text{ cm(o)}} \times \frac{1 \text{ cmHg}}{13.55 \text{ cm}H_2O} \times \frac{10 \text{ torr}}{1 \text{ cmHg}} = \underline{\underline{882 \text{ torr (B)}}}$$

6.22

	I	F
P	740 t	900 t
(V)	350 ml	? ml
T	—	—
N	—	—

$$? \text{ml(F)} = 350 \text{ ml(I)} \times \frac{740 \text{ t}}{900 \text{ t}} = \underline{\underline{288 \text{ ml(F)}}}$$

6.24

	I	F
(P)	475 t	? t
V	540 ml	320 ml
T	—	—
N	—	—

$$? \text{t(F)} = 475 \text{ t(I)} \times \frac{540 \text{ ml}}{320 \text{ ml}} = \underline{\underline{802 \text{ t(F)}}}$$

6.28

	I	F
P	—	—
(V)	2.00 ℓ	? ℓ
T	299 K	373 K
N	—	—

$$? \ell\text{(F)} = 2.00 \ \ell\text{(I)} \times \frac{373 \text{ K}}{299 \text{ K}} = \underline{\underline{2.49 \ \ell\text{(F)}}}$$

6.30

	I	F
P	1 atm	1 atm
V	0.400 ℓ	0.850 ℓ
(T)	305 K	? K
N	—	—

$$? \text{ K(F)} = 305 \text{ K(I)} \times \frac{0.850 \ \ell}{0.400 \ \ell} = \underline{\underline{648 \text{ K(F)}}}$$

6.34 At the initial conditions 1.00 ℓ of CO_2 weighs 1.96 g. The mass of 1.00 ℓ of CO_2 (FINAL) will be the density of CO_2 (FINAL)

	I	F
P	760 t	650 t
V	1.00 ℓ	1.00 ℓ
T	273 K	298 K
(N)	1.96 g	? g

$$? \text{g(F)} = 1.96 \text{ g(I)} \times \frac{650 \text{ t} \ 273 \text{ K}}{760 \text{ t} \ 298 \text{ K}} = \underline{\underline{1.54 \text{ g(F)}}}$$

(DENSITY g/ℓ)

6.38 $200t + 500t + 150t = \underline{\underline{850t}}$

6.40 I ⟶ F (N_2)

(P)	740 t	? t
V	20.0 ml	50.0 ml
T	273 K	273 K
N	———	———

$$? t(F) = 740 \cancel{t(I)} \times \frac{20.0 \cancel{ml}}{50.0 \cancel{ml}} = \underline{\underline{296\ t(F)\ (N_2)}}$$

I ⟶ F (O_2)

(P)	640 t	? t
V	30.0 ml	50.0 ml
T	273 K	273 K
N	———	———

$$? t(F) = 640 \cancel{t(I)} \times \frac{30.0 \cancel{ml}}{50.0 \cancel{ml}} = \underline{384\ t(F)\ (O_2)}$$

$$P_T = P_{N_2} + P_{O_2} = \underline{\underline{680\ t(F)}}$$

6.42 I ⟶ F (STP)

P	(700-24)	760 t
(V)	100 ml	? ml
T	298 K	273 K
N	———	———

$$? ml(F) = 100 \cancel{ml(I)} \times \frac{676 \cancel{t}\ 273 \cancel{K}}{760 \cancel{t}\ 298 \cancel{K}} = \underline{\underline{81.5\ ml(F)}}$$

6.46 $? g\,SO_2 = 0.245 \cancel{\ell\,SO_2(STP)} \times \frac{64.1\,g\,SO_2}{22.4 \cancel{\ell\,SO_2(STP)}} = \underline{\underline{0.701\,g\,SO_2}}$

6.48 I (STP) ⟶ F (S)

P	1 atm	1 atm
V	1 ℓ	22.4 ℓ
T	273 K	273 K
(N)	1.96 g	? g

M.W. = g/mol = g/22.4 ℓ (STP)

$\therefore$

$$? g(F) = 1.96 \cancel{g(I)} \times \frac{22.4 \cancel{\ell}}{1.00 \cancel{\ell}} = \underline{\underline{43.9\ g(F)}}$$

(← M.W.)

6.50 I → F Ⓢ

	I	F Ⓢ
P	760 t	760 t
V	250 ml	22,400 ml
T	298 K	273 K
Ⓝ	0.164 g	? g

$$? g(F) = 0.164\ \cancel{g(I)} \times \frac{22{,}400\ \cancel{ml}\ 298\ \cancel{K}}{250\ \cancel{ml}\ 273\ \cancel{K}} = \underline{\underline{16.0\ g(F)}}$$

6.52

$$? g(F) = 1.81\ \cancel{g(I)} \times \frac{22.4\ \cancel{l}\ 303\ \cancel{K}}{1.00\ \cancel{l}\ 273\ \cancel{K}} = \underline{\underline{45.0\ g(F)}}$$

	I	F Ⓢ
P	760 t	760 t
V	1.00 l	22.4 l
T	303 K	273 K
Ⓝ	1.81 g	? g

6.54 I(STP) → F Ⓢ

	I (STP)	F Ⓢ
P	760 t	760 t
V	500 ml	22,400 ml
T	273 K	273 K
Ⓝ	0.6695 g	? g

$$? g(F) = 0.6695\ \cancel{g(I)} \times \frac{22{,}400\ \cancel{ml}}{500\ \cancel{ml}} = \underline{\underline{30.0\ g(F)}}\ \ (b)$$

$$? mol\ C = 80.0\ \cancel{g\ C} \times \frac{1\ mol\ C}{12.0\ \cancel{g\ C}} = \underline{6.67\ mol\ C}$$

$$? mol\ H = 20.0\ \cancel{g\ H} \times \frac{1\ mol\ H}{1.008\ \cancel{g\ H}} = \underline{19.8\ mol\ H}$$

∴ (a) $\underline{\underline{CH_3}}$ (b) $\underline{\underline{C_2H_6}}$

6.56 (a)

$$? ml\ N_2(STP) = 1.40 \times 10^{-3}\ \cancel{mol\ NO} \times \frac{1\ \cancel{mol\ N_2}}{2\ \cancel{mol\ NO}} \times \frac{22{,}400\ ml\ N_2(STP)}{1\ \cancel{mol\ N_2}} = \underline{\underline{15.7\ ml\ N_2(STP)}}$$

$$(b)\ ? ml\ N_2(STP) = 1.3 \times 10^{-3}\ \cancel{g\ H_2} \times \frac{1\ \cancel{mol\ H_2}}{2.016\ \cancel{g\ H_2}} \times \frac{1\ \cancel{mol\ N_2}}{2\ \cancel{mol\ H_2}} \times \frac{22{,}400\ ml\ N_2(STP)}{1\ \cancel{mol\ N_2}} = \underline{\underline{7.2\ ml\ N_2\ (STP)}}$$

6.58

$$? mol\ NO_2 = 10.0\ \cancel{g\ HNO_3} \times \frac{1\ \cancel{mol\ HNO_3}}{63.0\ \cancel{g\ HNO_3}} \times \frac{3\ mol\ NO_2}{2\ \cancel{mol\ HNO_3}} = \underline{\underline{0.238\ mol\ NO_2}}$$

	I Ⓢ	F
P	760 t	770 t
V	22,400 ml	? ml
T	273 K	298 K
N	1 mol	0.238 mol

$$? ml(F) = 22{,}400\ \cancel{ml(I)} \times \frac{760\ \cancel{t}\ 298\ \cancel{K}\ 0.238\ \cancel{mol}}{770\ \cancel{t}\ 273\ \cancel{K}\ 1\ \cancel{mol}} = \underline{\underline{5740\ ml(F)}}$$

6.60 At temperature "T" $\overline{KE}_{CH_4} = \overline{KE}_{CO_2}$ $\overline{KE} = 1/2\, m\bar{v}^2$

$$\frac{1}{2} m_{CH_4} \bar{v}^2_{CH_4} = \frac{1}{2} m_{CO_2} \bar{v}^2_{CO_2} \qquad \bar{v}^2_{CO_2} = \frac{m_{CH_4} \bar{v}^2_{CH_4}}{m_{CO_2}}$$

$$\bar{v}_{CO_2} = \left(\frac{m_{CH_4}}{m_{CO_2}}\right)^{\frac{1}{2}} \bar{v}_{CH_4} \qquad \bar{v}_{CH_4} = 1000\text{ mph} \quad \therefore \bar{v}_{CO_2} = \left(\frac{16}{44}\right)^{\frac{1}{2}} \times 1000\text{ mph}$$

$$\therefore \bar{v}_{CO_2} = \underline{\underline{600\text{ mph}}}$$

6.62 How much more precise answer can be obtained by the use of van der Waals equation? $\left(P + \frac{a}{V^2}\right)(V - b) = RT$

$$P = \frac{RT}{V-b} - \frac{a}{V^2}$$

$$P = \frac{0.082056\ \ell\cdot\text{atm} \times 273.18\ K}{1\ \text{mol}\cdot 1\ K\,(22.400\ \ell\,\text{mol}^{-1} - 0.02370\ \ell\,\text{mol}^{-1})} \ (1.00175) - \frac{0.03412\ \ell^2\ \text{atm}\ \text{mol}^{-2}}{(22.400\ \ell\,\text{mol}^{-1})^2} \ (6.80006 \times 10^{-5}) = \underline{\underline{1.0017\ \text{atm}}}$$

For I.G.: $P = \underline{\underline{1.000\ \text{atm}}}$

$\therefore$ He behaves almost exactly as an I.G. under these conditions.

7.32 $n\lambda = 2d \sin\theta \quad \therefore d = n\lambda/2 \sin\theta$

$? \text{Å} = \frac{1 \times 1.41 \text{ Å}}{2 \times \sin 20°} = \underline{\underline{2.06 \text{ Å}}}$ ← d

for $\theta = 27.4°$ $\underline{\underline{d = 1.53 \text{ Å}}}$

for $\theta = 35.8°$ $\underline{\underline{d = 1.21 \text{ Å}}}$

7.34

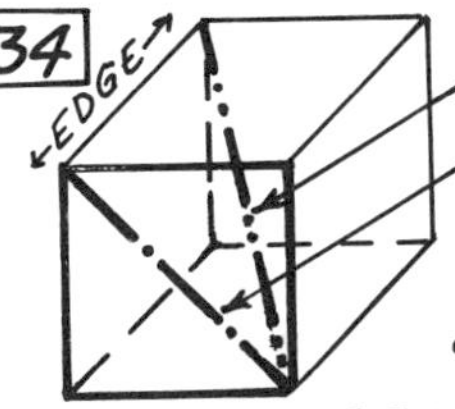

$(BD)^2 = (E)^2 + (FD)^2 \quad (FD)^2 = (E)^2 + (E)^2$

$\therefore (BD)^2 = 3(E)^2 \quad E = 2.884 \text{Å} \quad \therefore BD^2 = 3(2.884)^2$

$BD = 4.995 \text{ Å} \quad r = BD/4 = \underline{\underline{1.249 \text{ Å}}}$

7.36 for Face-Centered Cubic: Face-diagonal = 4 radii

$(FD)^2 = (E)^2 + (E)^2 = 2E^2 = (4r)^2 \quad \therefore \quad 4r = E(\sqrt{2}) \quad \underline{\underline{r = 1.4420 \text{Å}}}$

7.38 from 7.34: $BD = (3E^2)^{\frac{1}{2}} \quad BD = 2 \times r_{Cl^-} + 2 \times r_{Cs^+} \quad E = 4.123 \text{Å}$

$BD = 7.141 \text{ Å} \quad r_{Cl^-} = 1.81 \text{ Å} \quad \therefore \underline{\underline{r_{Cs^+}}} = [7.141 - 2 \times (1.81)] \text{Å}/2 = \underline{\underline{1.76 \text{ Å}}}$

7.40 Simple Cubic: $E = 2r$ from 7.34: (Body-Centered Cubic)

from 7.36: (FCC) $FD = 4r = (2E^2)^{\frac{1}{2}} \quad 4r = (3E^2)^{\frac{1}{2}}$

(a) $E = 2.88 \text{ Å}$, 1At/u.c.

$\frac{? \text{ gAg}}{1 \text{ cm}^3 \text{Ag}} = \frac{107.9 \text{ gAg}}{6.022 \times 10^{23} \text{ Ag At.}} \times \frac{1 \text{ Ag At.}}{1 \text{ u.c.}} \times \frac{1 \text{ u.c.}}{(2.88 \text{ Å})^3} \times \frac{(10^8 \text{ Å})^3}{(1 \text{ cm})^3} =$

$\underline{\underline{7.50 \text{ gAg/cm}^3\text{Ag}}}$

(b) $E = 3.33 \text{ Å}$, 2At/u.c.

$\frac{? \text{ gAg}}{1 \text{ cm}^3 \text{Ag}} = \frac{107.9 \text{ gAg}}{6.022 \times 10^{23} \text{ At.}} \times \frac{2 \text{ At.}}{1 \text{ u.c.}} \times \frac{1 \text{ u.c.}}{(3.33 \text{ Å})^3} \times \frac{(10^8 \text{ Å})^3}{1 \text{ cm}^3} = \underline{\underline{9.70}} \frac{\text{gAg}}{\text{cm}^3 \text{Ag}}$

(c) $E = 4.07 \text{ Å}$, 4 At./u.c.

$\frac{? \text{gAg}}{1 \text{ cm}^3 \text{Ag}} = \frac{107.9 \text{ gAg}}{6.022 \times 10^{23} \text{ At.}} \times \frac{4 \text{ At.}}{(4.07 \text{ Å})^3} \times \frac{(10^8 \text{ Å})^3}{1 \text{ cm}^3} = \underline{\underline{10.6 \frac{\text{gAg}}{\text{cm}^3 \text{Ag}}}}$

8.42 $? \text{kcal} = 55.0\,\text{g(A)} \times \frac{9.22\,\text{kcal}}{1\,\text{mol(A)}} \times \frac{1\,\text{mol(A)}}{46.07\,\text{g(A)}} = \underline{\underline{11.0\,\text{kcal}}}$

8.44 $? \text{kJ} = 1\,\text{mol Hg} \times \frac{200.6\,\text{g Hg}}{1\,\text{mol Hg}} \times \frac{4.29\,\text{kJ}}{14.5\,\text{g Hg}} = \underline{\underline{59.3\,\text{kJ}}}$

8.50 $\ln\left(\frac{P_1}{P_2}\right) = \frac{-\Delta H_{VAP}}{R}\left(\frac{1}{T_1} - \frac{1}{T_2}\right)$ $\quad ? \text{atm} = 185\,\text{t} \times \frac{1\,\text{atm}}{760\,\text{t}} = \underline{0.243\,\text{atm}}$

$\Delta H_{VAP} = -R \ln\frac{P_1}{P_2} \times \frac{T_1 T_2}{T_2 - T_1} = -1.99\,\text{cal mol}^{-1}\,{}^\circ\text{K}^{-1} \ln\left(\frac{0.243}{0.384}\right) \times \frac{273 \times 283\,({}^\circ\text{K})^2}{(283-273)\,{}^\circ\text{K}}$

$= \underline{\underline{7.04 \times 10^3\,\text{cal mol}^{-1}}}$ $\quad$ or $\quad \underline{\underline{29.4\,\text{kJ mol}^{-1}}}$

9.26 $\frac{?\text{ mol Gly.}}{1\text{ mol Sol.}} = \frac{45.0\text{ g Gly.}}{(45.0/92.0 + 100.0/18)\text{ mol Sol.}} \times \frac{1.00\text{ mol Gly.}}{92.0\text{ g Gly.}} = 8.09 \times 10^{-2}$ (Mole Fract.)

$\frac{?\text{ g Gly.}}{100.0\text{ g Sol}} = \frac{45.0\text{ g Gly.}}{(45.0 + 100.0)\text{ g Sol.}} \times \frac{100.0\text{ g Sol.}}{100.0\text{ g Sol.}} = 31.0\%$ (Gly. Sol.)

$\frac{?\text{ mol Gly.}}{1.00\text{ Kg }H_2O} = \frac{45.0\text{ g Gly}}{100.0\text{ g }H_2O} \times \frac{1000\text{ g }H_2O}{1.00\text{ kg }H_2O} \times \frac{1\text{ mol Gly.}}{92.0\text{ g Gly.}} = 4.89$ molal

9.28 (a) $\frac{?\text{ g }Zn(NO_3)_2}{100\text{ g Sol.}} = \frac{121.8\text{ g }Zn(NO_3)_2}{1000\text{ ml Sol.}} \times \frac{1\text{ ml Sol.}}{1.107\text{ g Sol.}} \times \frac{100\text{ g Sol.}}{100\text{ g Sol.}} = 11.0\%$ ($Zn(NO_3)_2$ Sol.)

(b) $\frac{?\text{ mol }Zn(NO_3)_2}{1\text{ Kg }H_2O} = \frac{11.0\text{ g }Zn(NO_3)_2}{89.0\text{ g }H_2O} \times \frac{1000\text{ g }H_2O}{1\text{ Kg }H_2O} \times \frac{1\text{ mol }Zn(NO_3)_2}{189.4\text{ g }Zn(NO_3)_2} = 0.653\text{ m}$

(c) $\frac{?\text{ mol }Zn(NO_3)_2}{1\text{ mol Sol.}} = \frac{11.0/189.4\text{ mol }Zn(NO_3)_2}{(11.0/189.4 + 89.0/18.0)\text{ mol Sol}} = 0.0116$ (mole fract)

(d) $\frac{?\text{ mol }Zn(NO_3)_2}{1\ l\text{ sol.}} = \frac{121.8\text{ g }Zn(NO_3)_2}{1\ l\text{ sol.}} \times \frac{1\text{ mol }Zn(NO_3)_2}{189.4\text{ g }Zn(NO_3)_2} = 0.643$ MOLAR

9.30 $\frac{?\text{ mol E.G.}}{1\text{ kg }H_2O} = \frac{222.6\text{ g E.G.}}{200\text{ g }H_2O} \times \frac{1\text{ mol E.G.}}{62.0\text{ g E.G.}} \times \frac{1000\text{ g }H_2O}{1\text{ kg }H_2O} = 18.0\text{ m}$

$\frac{?\text{ mol E.G.}}{1\ l\text{ sol.}} = \frac{222.6\text{ g E.G.}}{(222.6 + 200)\text{ g sol.}} \times \frac{1\text{ mol E.G.}}{62.0\text{ g E.G.}} \times \frac{1072\text{ g sol.}}{1\ l\text{ sol.}} = 9.11\text{ M}$

9.32 $\frac{?\text{ g Alc.}}{100\text{ g Sol.}} = \frac{0.250 \times 60.0\text{ g Alc.}}{(0.250 \times 60.0 + 0.750 \times 18.0)\text{ g sol.}} \times \frac{100\text{ g Sol.}}{100\text{ g Sol.}} = 0.526\%$ (Alc.)

9.32 cont

$$\frac{?\,\text{mol Alc.}}{1\text{ kg }H_2O} = \frac{0.250\,\text{mol Alc.}}{0.750\,\text{mol }H_2O} \times \frac{1\text{ mol }H_2O}{18.0\text{ g }H_2O} \times \frac{1000\text{ g }H_2O}{1\text{ kg }H_2O} = 18.5\ m$$

9.34

$$\frac{?\,\text{mol NaCl}}{1\text{ mol Sol.}} = \frac{6.25\,\text{mol NaCl}}{1\text{ kg }H_2O} \times \frac{1\text{ kg }H_2O}{(1000/18.0 + 6.25)\text{ mol Sol.}} = 0.101 \quad \text{(MOLE FRACT.)}$$

$$\frac{?\,\text{g NaCl}}{1\text{ g Sol.}} = \frac{6.25\,\text{mol NaCl}}{1\text{ kg }H_2O} \times \frac{58.45\text{ g NaCl}}{1\text{ mol NaCl}} \times \frac{1\text{ kg }H_2O}{(1000 + 6.25 \times 58.45)\text{ g Sol.}} = 0.268 \quad \text{(WT. FRACT.)}$$

9.36

$$?\,\text{cal} = 10.0\text{ g }AlCl_3 \times \frac{1\text{ mol }AlCl_3}{133.3\text{ g }AlCl_3} \times \frac{-76.8 \times 10^3\text{ cal}}{1\text{ mol }AlCl_3} = -5760\text{ cal}$$

9.40

$$\frac{?\,\text{t Eth.}}{5.00 \times 10^{-2}\text{ g Eth.}} = \frac{751\text{ t Eth.}}{6.56 \times 10^{-2}\text{ g Eth.}} \times \frac{5.00 \times 10^{-2}\text{ g Eth.}}{5.00 \times 10^{-2}\text{ g Eth.}} = \frac{572\text{ t Eth.}}{5.00 \times 10^{-2}\text{ g Eth.}}$$

9.42

$$?\,t\,(\text{sol}) = 93.4\,t \times \frac{1\text{ mol Sol.}}{1\text{ mol }B_3.} \times \overbrace{\frac{1000/78.0\text{ mol }B_3.}{(1000/78.0 + 56.4/282)\text{ mol Sol.}}}^{\text{mole fraction }B_3} = 92.0\,t\,(\text{sol})$$

9.46

$$\frac{?\,\text{g X}}{1\text{ mol X}} = \frac{1\text{ kg }B_3 \times 4.9\,C°}{1\text{ mol X}} \times \frac{3.84\text{ g X}}{0.500\text{ kg }B_3 \times 0.307\,C°} = \frac{120\text{ g X}}{1\text{ mol X}}$$

$\therefore$ M.W. $= 2 \times 64.0 = 128$ FORMULA $= C_8H_4N_2$

9.48

$$?\,\text{g G.} = 0.750\,C° \times 150\text{ g }H_2O \times \frac{1\text{ mol G.}}{1\text{ kg }H_2O \times 1.86\,C°} \times \frac{180\text{ g G.}}{1\text{ mol G.}} \times \frac{1\text{ kg }H_2O}{10^3\text{ g }H_2O} =$$

10.9 g G.

$$?\,C°(\Delta BP) = -0.750\,C°(\Delta f.P.) \times \frac{0.512\,C°(\Delta BP)}{-1.86\,C°(\Delta f.p.)} = 0.206\,(\Delta BP)$$

$\therefore$ BP $= 100.206\,°C$

10.24 $W_{max} = RT \ln P_1/P_2$ (for 1 mol I.G.: $PV=RT$)

$\therefore W_{max} = PV \ln P_1/P_2 = 15.0 \text{ atm} \times 10.0 \ell \times \ln 15/1 = \underline{\underline{406 \ \ell\cdot\text{atm}}}$
$\underline{\underline{(4.11\times10^4 J)}}$

10.26 $CaO_{(s)} + H_2O_{(\ell)} \rightarrow Ca(OH)_{2(s)}$ (298°K) (1 atm)

$q_P = -15.6$ kcal $= \Delta H$ $\quad \Delta E = \Delta H - \Delta(PV)$ @ constant P:

$\Delta(PV) = P\Delta V$, $\Delta V = V_f - V_i$ $\quad \Delta V = 74.10g \times \frac{1.00ml}{2.24g} - 56.08g \times \frac{1 ml}{3.45g} - 18.02g \times \frac{1ml}{0.997g} = \underline{-1.25 ml}$

$\therefore P\Delta V = \underline{-1.25 ml\cdot atm} = \underline{-3.02\times10^{-5} kcal}$

Since $\Delta H = \underline{-15.6 \text{ kcal mol}^{-1}}$ the $P\Delta V$ correction is smaller than the uncertainty of the ΔH value! $\therefore \underline{\underline{\Delta H \approx \Delta E}}$ (all substances liq. or solid)

10.30 $\Delta H_{vap.}(H_2O)$ @ 25°C $= \underline{10.5 \text{ kcal mol}^{-1}}$ $\quad \Delta(PV) = P(V_{vap.} - V_{liq}) \approx PV_{vap.}$ (assume I.G.) $PV_{vap.} = RT = 0.592 \text{ kcal mol}^{-1}$

$\therefore \Delta E_{vap.} = (10.5 - 0.592) = \underline{\underline{9.9 \text{ kcal mol}^{-1}}}$

$q_P = \Delta H = \underline{\underline{10.5 \text{ kcal mol}^{-1}}}$ $\quad W = P\Delta V = \underline{\underline{0.592 \text{ kcal mol}^{-1}}}$

10.32 (a) $C_2H_{2(g)} + H_{2(g)} \rightarrow C_2H_{4(g)}$ $\quad \Delta H° = \Delta H_f°(C_2H_4) - \Delta H_f°(C_2H_2) - \Delta H_f°(H_2) = 12.4 \text{ kcal} - 54.2 \text{ kcal} - 0 \text{ kcal} = \underline{\underline{-41.8 \text{ kcal mol}^{-1}}}$

(b) $\underline{\underline{-31.6 \text{ kcal mol}^{-1}}}$ (c) $\underline{\underline{-40.8 \text{ kcal}}}$ (d) $\underline{\underline{9.9 \text{ kcal mol}^{-1}}}$

(e) $\underline{\underline{-683.7 \text{ kcal}}}$

10.34 $\Delta H_f°(FeO) = ?$ $\quad Fe_{(s)} + \frac{1}{2}O_{2(g)} \rightarrow FeO_{(s)}$

from 10.33:

$Fe_{(s)} + \cancel{CO_{2(g)}} \rightarrow FeO_{(s)} + \cancel{CO_{(g)}}$	$\Delta H = 4.00 \text{ kcal mol}^{-1}$
$\cancel{C_{(s)}} + O_{2(g)} \rightarrow \cancel{CO_{2(g)}}$	$\Delta H = -94.1 \text{ kcal mol}^{-1}$
$\cancel{CO_{(g)}} \rightarrow \cancel{C_{(s)}} + \frac{1}{2}O_{2(g)}$	$\Delta H = 26.4 \text{ kcal mol}^{-1}$
$Fe_{(s)} + \frac{1}{2}O_{2(g)} \rightarrow FeO_{(s)}$	$\underline{\underline{\Delta H_f°(FeO) = -63.7 \text{ kcal/mol}}}$

10.36 $O_{3(g)} + O_{(g)} \rightarrow 2O_{2(g)} \leftarrow \left\{ \begin{array}{l} O_{3(g)} + Cl_{(g)} \rightarrow O_{2(g)} + ClO_{(g)} \\ ClO_{(g)} + O_{(g)} \rightarrow Cl_{(g)} + O_{2(g)} \end{array} \right\}$

$\therefore \Delta H^\circ = \Delta H_1 + \Delta H_2 = (-30-64)\text{ kcal}$ $\quad \underline{\underline{\Delta H^\circ = -94 \text{ kcal}}}$

10.38 (a) $2NO_{(g)} + O_{2(g)} \rightarrow 2NO_{2(g)}$ $\quad \Delta H^\circ = \{2(8.09) - 0 - 2(21.6)\}\text{ kcal}$

$\Delta H^\circ = \underline{\underline{-27.0 \text{ kcal}}} \quad = \underline{\underline{-113 \text{ kJ}}}$ $\quad$ (b) $\underline{\underline{73.1 \text{ kcal} \quad (306 \text{ kJ})}}$

(c) $\underline{\underline{-25.5 \text{ kcal} \quad (-107 \text{ kJ})}}$

10.40 (1) $C_2H_5OH + \frac{1}{2}O_2 \rightarrow CH_3CHO + H_2O$ $\quad \Delta H_1^\circ = -39.8 - 68.3 + 66.4 + 0$

(2) $CH_3CHO + \frac{1}{2}O_2 \rightarrow CH_3COOH$ $\quad \Delta H_2^\circ = -116.4 + 39.8 + 0$

(3) $CH_3COOH + 2O_2 \rightarrow 2CO_2 + 2H_2O$ $\quad \Delta H_3^\circ = -2(94.1) - 2(68.3) + 116.4$

Net Eq.: $\underline{\underline{C_2H_5OH + 3O_2 \rightarrow 2CO_2 + 3H_2O}}$ $\quad \Delta H_1^\circ = -41.7, \Delta H_2^\circ = -76.6, \Delta H_3^\circ = -208$

$\underline{\underline{\Delta H_T^\circ = -327 \text{ kcal}}}$

10.42

$$? \text{g Carb.} = 2000 \text{ kcal} \times \frac{1 \text{ g Carb.}}{4 \text{ kcal}} = \underline{\underline{500 \text{ g Carb.}}}$$

10.52 $\Delta H_{fus}(H_2O)$ @ 0°C $= 1.44$ kcal mol^{-1} $\quad \Delta H_{vap}$ (100°C) $= 9.72$ kcal

$\Delta S = \frac{q_{rev}}{T} = \frac{\Delta H}{T}$ $\quad (\Delta H = q_P$ and these processes are @ 1 atm)

$$\therefore \Delta S_{vap} = \frac{9720 \text{ cal mol}^{-1}}{373 \text{ K}} = \underline{\underline{26.1 \text{ cal mol}^{-1}\text{K}^{-1}}}$$

$$\Delta S_{fus} = \frac{1440 \text{ cal mol}^{-1}}{273 \text{ K}} = \underline{\underline{5.27 \text{ cal mol}^{-1}\text{K}^{-1}}}$$

Both melting & vaporizing involve increases in randomness ($\Delta S = +$). Vaporizing also involves a large volume increase ($\Delta S =$ LARGE $+$).

10.54 (a) net $-\frac{1}{2}$ mol gas (b) same $\Delta S_{(a)} = 61.3\,eu - \{59.3 + \frac{1}{2}(49.0)\}\,eu$

$\Delta S_{(a)} = -22.5\,eu$ $\Delta S_{(b)} = 51.06\,eu - \{47.3 + \frac{1}{2}(49.0)\}\,eu = -20.7\,eu$

$\therefore$ ans $\rightarrow$ (a)

10.58 $\Delta G^\circ = [3\times(-94.3) + 4(-54.6) - (-5.6) - 5(0)]$ kcal $= -495.7$ kcal

(-94.3 = $\Delta G_f^\circ(CO_2)$; -54.6 = $\Delta G_f^\circ(H_2O)(g)$; -5.6 = $\Delta G_f^\circ(C_3H_8)(g)$; 0 = $\Delta G_f^\circ(O_2)$)

ΔG° = max useful work. Since any real process can only approach reversible conditions, actual work $< \Delta G^\circ$

CHAPTER 11

11.32 $2A \rightarrow 4B + C$

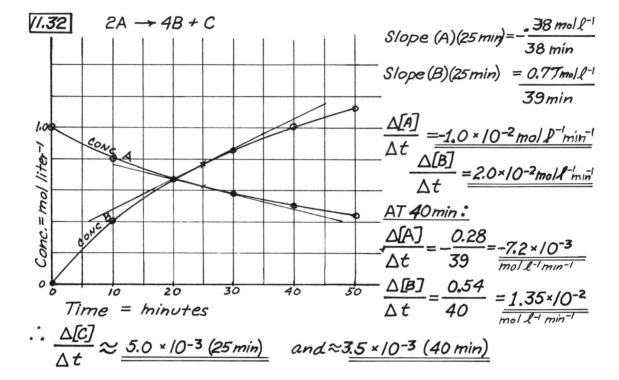

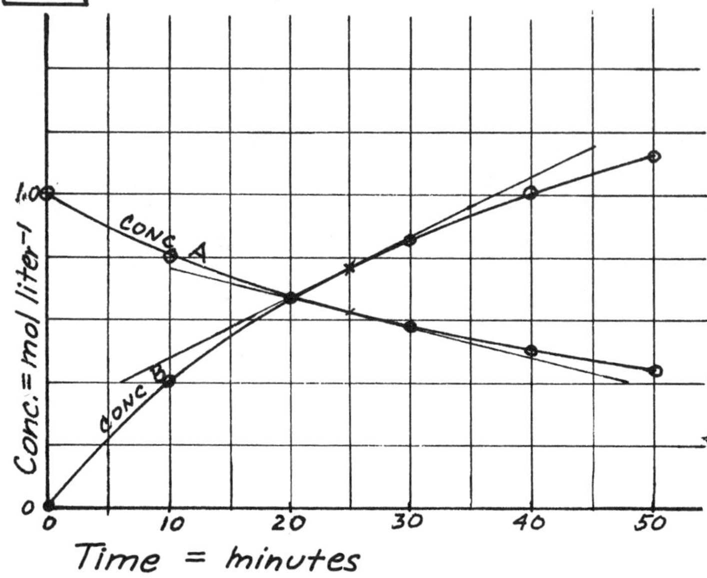

$$\text{Slope (A)(25 min)} = -\frac{.38 \text{ mol } \ell^{-1}}{38 \text{ min}}$$

$$\text{Slope (B)(25 min)} = \frac{0.77 \text{ mol } \ell^{-1}}{39 \text{ min}}$$

$$\frac{\Delta[A]}{\Delta t} = -1.0 \times 10^{-2} \text{ mol } \ell^{-1} \text{ min}^{-1}$$

$$\frac{\Delta[B]}{\Delta t} = 2.0 \times 10^{-2} \text{ mol } \ell^{-1} \text{ min}^{-1}$$

AT 40 min:

$$\frac{\Delta[A]}{\Delta t} = -\frac{0.28}{39} = -7.2 \times 10^{-3} \text{ mol } \ell^{-1} \text{ min}^{-1}$$

$$\frac{\Delta[B]}{\Delta t} = \frac{0.54}{40} = 1.35 \times 10^{-2} \text{ mol } \ell^{-1} \text{ min}^{-1}$$

$$\therefore \frac{\Delta[C]}{\Delta t} \approx 5.0 \times 10^{-3} \text{ (25 min)} \quad \text{and} \approx 3.5 \times 10^{-3} \text{ (40 min)}$$

11.34 $k = 1.63 \times 10^{-1}$ liter mol^{-1} sec^{-1} $\quad$ Rate $= k[ICl][H_2]$

(a) Rate $= 1.63 \times 10^{-1}$ ~~liter mol^{-1}~~ sec^{-1} (0.25 ~~mol liter^{-1}~~)(0.25 mol liter^{-1})

$= 1.0 \times 10^{-2}$ mol liter^{-1} sec^{-1} $\quad$ (b) $= 2.0 \times 10^{-2}$ $\quad$ (c) $= 4.1 \times 10^{-2}$

11.36 $2\,NOCl \rightarrow 2\,NO + Cl_2$ $\quad$ Rate $= k[NOCl]^X$

$2 \times [NOCl] \rightarrow 4 \times$ Rate $\quad$ $3 \times [NOCl] \rightarrow 9 \times$ Rate $\quad \therefore X = 2$

(a) Rate $= k[NOCl]^2$ $\quad$ (b) $k = \dfrac{3.60 \times 10^{-9} \text{ mol liter}^{-1} \text{ sec}^{-1}}{(0.30 \text{ mol liter}^{-1})^2} = 4.0 \times 10^{-8} \ \ell \text{ mol}^{-1} \text{ s}^{-1}$

(c) $(1.5)^2 = 2.25$

11.38 $k(230°C) = \underline{0.163\ \ell^2 mol^{-2} s^{-1}}$ $k(240°C) = \underline{0.348}$

$k(230°C) = A e^{-Ea/503°R}$ $k(240°C) = A e^{-Ea/513°R}$

$\therefore \frac{0.163}{0.348} = \frac{A e^{-Ea/503°R}}{A e^{-Ea/513°R}} = e^{-Ea/503°R}\ e^{Ea/513°R}$

$0.468 = e^{503Ea - 513Ea/503 \cdot 513R}$ $\ln 0.468 = -10Ea/5.13 \times 10^5$

$\underline{\underline{Ea}} = 0.759 \times 5.13 \times 10^5/10 = \underline{\underline{3.9 \times 10^4\ cal\ mol^{-1}}} = \underline{\underline{39\ kcal\ mol^{-1}}}$

$A = k(230°C)/e^{-Ea/RT} = 0.163\ e^{Ea/RT} = \underline{\underline{1.36 \times 10^{16}\ \ell^2 mol^{-2} s^{-1}}}$

11.40 $Ea = 182\ kJ\ mol^{-1}$ $k(700°C) = 1.57 \times 10^{-3}\ \ell\ mol^{-1} sec^{-1}$

$k(600°C) = A e^{-Ea/RT}$ $A = k(700°C)\ e^{Ea/RT}$

$\therefore k(600°C) = k(700°C) e^{Ea/R973} \times e^{-Ea/R873}$

$k(600°C) = 1.57 \times 10^{-3}\ \ell\ mol^{-1} sec^{-1}\ e^{873Ea - 973Ea/R \times 973 \times 873}$

$= \underline{\underline{1.19 \times 10^{-4}\ \ell\ mol^{-1} sec^{-1}}}$

CHAPTER 12

12.22 $K_P = \frac{P_{PCl_3}\ P_{Cl_2}}{P_{PCl_5}} = \frac{(0.23)(0.055)}{(0.0023)} = 5.5 \text{ atm}$ (1)

5.55 (2) 5.43 (3) 5.49 (4) (Within experimental precision all of these represent a $K_P = 5.5$ atm.)

$K_C = K_P (RT)^{-\Delta n} = 5.5 \text{ atm } (0.08205\ \ell\text{-atm deg}^{-1}\text{mol}^{-1}\ 298°K)^{-\Delta n}$

$\Delta n = +1$ ∴ $K_C = 0.22 \text{ mol } \ell^{-1}$

12.24 $\Delta G° = -RT \ln K_P$ $K_P = e^{-\Delta G°/RT}$ $K_P = e^{+3{,}220 \text{ cal mol}^{-1}/1.987 \cdot 700 \text{ (cal mol}^{-1})}$

$K_P = 10.13$

12.26 $\Delta G° = -RT \ln K_P = -1.987 \text{ cal mol}^{-1}\text{deg}^{-1} \times 800°K \ln 5.10$

$\Delta G°(800°K) = -2{,}590 \text{ cal mol}^{-1}$

12.30 Eq 12.2: $\Delta G = \Delta G° + 2.303 RT \log Q$ AT EQUILIBRIUM $\Delta G = 0$

∴ $\Delta G° = -2.303 RT \log Q = -3{,}220 \text{ cal mol}^{-1}$ $\log Q = 1.004$

$Q = 10.09$ (for equilibrium)(K_{EQ}) $Q = \frac{P_{CH_3OH}}{P_{H_2}^2 \times P_{CO}} = \frac{3 \times 10^{-6} \text{ atm}}{(1 \times 10^{-2} \text{ atm})^2 (2 \times 10^{-3} \text{ atm})}$

$Q(calc) = 15 \neq Q_{EQ}\ (K_{EQ})$

∴ System is not at equilibrium; $Q_{calc} > Q_{EQ}\ (K_{EQ})$; P_{CH_3OH} must decrease to reach equilibrium; Reaction: L ← R

12.32 (a) $\Delta G° = [(-184.9) + 6(-54.6) - (-531.0)] \text{ kcal mol}^{-1} = 18.5 \text{ kcal mol}^{-1}$

(b) $K_{EQ} = K_P = e^{-\Delta G°/RT} = e^{-31.24} = 2.70 \times 10^{-14}$

(c) $P_{(H_2O)(EQ)} = ?$ $K_P = (P_{H_2O})^6 = 2.70 \times 10^{-14}$ ∴ $P_{H_2O}(EQ) = 5.48 \times 10^{-3} \text{ atm}$

12.34 $K_p = \frac{P_{NOCl}^2}{P_{Cl_2} \; P_{NO}^2} = \frac{(0.15)^2}{(0.18)(0.65)^2} = \underline{\underline{0.30 \text{ atm}^{-1}}}$

$K_c = K_p(RT)^{-\Delta n}$ $\Delta n = -1$ $\therefore K_c = \frac{0.30 \text{ atm}^{-1}(0.08205 \, \ell \, \text{atm} \, °K^{-1} \times T)^{+1}}{1 \text{ mol}}$

$K_c = \underline{\underline{2.4 \times 10^{-2} (T) \, \ell \, mol^{-1}}}$

12.36 $K_c = \frac{[CO]^2[O_2]}{[CO_2]^2} = 6.4 \times 10^{-7}$ (2000°C)

let X = moles CO_2 decomposed

$\therefore [CO_2] = 1 \times 10^{-3} - X$

$[O_2] = X/2$, $[CO] = X$ $\therefore$ $6.4 \times 10^{-7} = \frac{X^2 \times 0.5X}{(1 \times 10^{-3} - X)^2}$ Assume: $X \ll 10^{-3}$

$\therefore [CO_2] \approx 10^{-3}$ and: $0.5X^3 = 6.4 \times 10^{-13}$ $X^3 = 12.8 \times 10^{-13}$

$X = 1.1 \times 10^{-4}$

(X is about a tenth of 10^{-3} so the assumption should give a fair value for the concentrations.)

$\therefore$ (a) $[CO] = 1.1 \times 10^{-4} \text{ mol } \ell^{-1}$, $[O_2] = 5.4 \times 10^{-5} \text{ mol } \ell^{-1}$

(b) $[CO_2] = 8.9 \times 10^{-4} \text{ mol } \ell^{-1}$ ← If this value (more accurate than 10^{-3}) is used for the $[CO_2]$, a more accurate value of X (1×10^{-4}) can be calc.

<u>BUT NOTE:</u> to 1 sig. fig. $\underline{\underline{[CO_2] = 1 \times 10^{-3}}}$, $\underline{\underline{[CO] = 1 \times 10^{-4}}}$, $\underline{\underline{[O_2] = 5 \times 10^{-5}}}$

12.38 $CO + Cl_2 = COCl_2$ $K_{EQ} = \frac{[COCl_2]}{[CO][Cl_2]} = 4.6 \times 10^9 \, \ell \, mol^{-1}$

let X = no. of moles $COCl_2$ decomp.

$\therefore \underline{[COCl_2]} = (0.20 - X) \text{ mol}/10.0 \, \ell = \underline{(0.020 - 0.100X) \text{ mol } \ell^{-1}}$ $[CO] = X/10.0 \text{ mol } \ell^{-1}$

$[Cl_2] = [CO]$ $\therefore 4.6 \times 10^9 = \frac{0.020 - 0.100X}{(0.100X)^2}$ Since K_c is so large, very little $COCl_2$ decomposes

($X \ll 0.20$) then $0.020 - 0.100X \approx 0.020$ $\therefore 4.6 \times 10^9 = \frac{0.020}{0.0100X^2}$

$X^2 = 4.35 \times 10^{-10}$, $\underline{X = 2.09 \times 10^{-5}}$

Note: 2.09×10^{-5} $\underline{\underline{\text{is}}}$ $\ll 0.20$ $\therefore$ $\underline{\underline{[COCl_2] = 0.02 , [CO] = [Cl_2] = 2.09 \times 10^{-6}}}$

14.16 (a) 10^{-3} M HCl $\rightarrow [H^+] = 10^{-3}$ $\therefore [OH^-] = 10^{-11}$ pH = 3.0

(c) 3.1×10^{-3} M NaOH $\rightarrow [OH^-] = 3.1 \times 10^{-3}$ $\therefore [H^+] = 3.2 \times 10^{-12}$ pH = 11.49

(d) 1.2×10^{-2} M $Ba(OH)_2$ —(assume complete dissoc.)$\rightarrow [OH^-] = 2.4 \times 10^{-2}$

$\therefore [H^+] = 4.2 \times 10^{-13}$
pH = 12.38

14.20 $pK_b = -\log K_b$ $\therefore$ $K_b = {}^{-pK_b} = 1.45 \times 10^{-4}$

14.22 (a) $NH_3 + H_2O = NH_4^+ + OH^-$ $K_b = \dfrac{[NH_4^+][OH^-]}{[NH_3]} = 1.8 \times 10^{-5}$

	H_2O	ADD	ADJ.	EQ	ASSUME:	$\therefore$
NH_3	—	0.15	$-X$	$0.15-X$	$X \ll 0.15$	0.15
NH_4^+	—	—	$+X$	X		X
OH^-	10^{-7}	—	$+X$	$10^{-7}+X$	$X \gg 10^{-7}$	X

$1.8 \times 10^{-5} = \dfrac{(X)(X)}{0.15}$

$\therefore X^2 = 2.7 \times 10^{-6}$ $X = 1.6 \times 10^{-3}$ NOTE: $0.15 \gg 0.0016 \gg 0.0000001$

$\therefore$ Both assumptions valid $[OH^-] = X = 1.6 \times 10^{-3}$

(b) $[OH^-] = 5.8 \times 10^{-4}$ (c) 1.7×10^{-2} (d) 6.2×10^{-5} (e) 4.2×10^{-6}

14.26 $HA = H^+ + A^-$ $K_a = \dfrac{[H^+][A^-]}{[HA]}$ $[HA] \approx 10^{-1}$ pH = 5.37

$\therefore [H^+] = 4.27 \times 10^{-6} \approx [A^-]$

$K_a = (4.27 \times 10^{-6})^2 / 10^{-1} = 1.8 \times 10^{-10}$

14.28 (a) 1.0 M HCO_2H $HCO_2H = H^+ + CO_2H^-$ $K_a = \dfrac{[H^+][CO_2H]}{[HCO_2H]}$

$K_a = 1.8 \times 10^{-4}$ (table 14.1)

	H_2O	ADD	ADJ.	EQ	ASSUME:	$\therefore$
HCO_2H	—	1.0	$-X$	$1.0-X$	$X \ll 1.0$	1.0
H^+	10^{-7}	—	$+X$	$10^{-7}+X$		X
CO_2H^-	—	—	$+X$	X	$X \gg 10^{-7}$	X

$1.8 \times 10^{-4} = \dfrac{(X)(X)}{1.0}$

$\therefore X^2 = 1.8 \times 10^{-4}$, $X = 1.3 \times 10^{-2}$ % ioniz. = $100 \times X/1.0$ = 1.3%

(b) 3.7% (c) 0.014% (d) 0.63% (e) 0.025% (f) assume 100%

14.30 (a) $1.8 \times 10^{-5} = \frac{[H^+][Ac^-]}{[HAc]}$ $[H^+] = \frac{0.25}{0.15} \times 1.8 \times 10^{-5} = \underline{\underline{3.0 \times 10^{-5}}}$

(c) $4.5 \times 10^{-4} = \frac{[H^+][NO_2^-]}{[HNO_2]}$ $[H^+] = \frac{0.30}{0.40} \times 4.5 \times 10^{-4} = \underline{\underline{3.4 \times 10^{-4}}}$

(b) $\underline{\underline{1.8 \times 10^{-4}}}$, (d) $[OH^-] = \underline{\underline{3.0 \times 10^{-5}}}$ $\therefore [H^+] = \underline{\underline{3.3 \times 10^{-10}}}$ (e) $\underline{\underline{9.8 \times 10^{-9}}}$

14.32 $HAc = H^+ + Ac^-$ $K_a = 1.8 \times 10^{-5}$ $K_a = \frac{[H^+][Ac^-]}{[HAc]}$

Conc	H_2O	ADD	ADD(2)	ADJ.	EQ
HAc	—	—	—	+Y	Y
H^+	10^{-7}	—	X	−Y	X−Y
Ac^-	—	1.0	—	−Y	1.0−Y

$pH = 4.74 \therefore [H^+] = 1.8 \times 10^{-5}$

$\therefore X - Y = 1.8 \times 10^{-5}$

$1.8 \times 10^{-5} = 1.8 \times 10^{-5} \times (1.0 - Y)/Y$ $\therefore Y = 1.0 - Y$ $Y = 0.5$

$X - Y = 1.8 \times 10^{-5}$, $X - 0.5 = 1.8 \times 10^{-5}$, $X = (50000 + 1.8) \times 10^{-5} \approx 0.5$

since X = molarity of the solution in HCl (0.5), 500 ml contain 0.25 mol HCl $? g\,HCl = 0.25 \text{ mol HCl} \times 36.5\,g\,HCl / 1.0 \text{ mol HCl}$

$\underline{\underline{9.1\ g\,HCl}}$

14.34 $\frac{? mol (c)}{1\,l} = \frac{500\,mg(c)}{200\,ml} \times \frac{1.00\,mol(c)}{1.76 \times 10^5\,mg(c)} \times \frac{10^3\,ml}{1.00\,l} = 1.42 \times 10^{-2}\,M(c)$

$K_{a1} = 7.94 \times 10^{-5}$

CONC	H_2O	ADD	ADD(2)	ADJ.	Assume:	EQ.
H(C)	—	—	0.0142	−X	$X \ll 10^{-2}$	0.0142
H^+	10^{-7}	10^{-1}	—	+X		10^{-1}
$(C)^-$	—	—	—	+X		X

$K_{a1} = \frac{[H^+][C^-]}{[H(C)]}$

$7.94 \times 10^{-5} = \frac{(10^{-1})\,X}{0.0142} =$ $X = 1.13 \times 10^{-5}$ (NOTE 10^{-5} is $\ll 10^{-2}$)

(dissociated)

% dissoc. $= \frac{1.13 \times 10^{-5}}{1.42 \times 10^{-4}} = \underline{\underline{0.0796\,\%}}$ (fraction = 0.000796)

14.36 $K_a = 1.4\times10^{-5} = \frac{[H^+][Nic.]}{[HNic.]} = \frac{(X)(X)}{(10^{-2}-X)}$ Assume: $X \ll 10^{-1}$ $\therefore 10^{-2}X \approx 10^{-2}$

$X^2 = 1.4\times10^{-7}$ $X = \underline{3.7\times10^{-4}}$ (3.7 is $\ll$ 100) $[H^+] = 3.7\times10^{-}$, $\underline{\underline{pH = 3.43}}$

14.38 $N_2H_4 + H_2O = N_2H_5^+ + OH^-$ $K_b = \frac{[N_2H_5^+][OH^-]}{[N_2H_4]} = 1.7\times10^{-6}$

$pH = 10.64$ $\therefore pOH = 3.36$, $[OH^-] \approx [N_2H_5^+]$

$[OH^-] = 4.36\times10^{-4}$ $[N_2H_4] = [N_2H_5^+][OH^-]/K_b = \frac{(4.36\times10^{-4})^2}{1.7\times10^{-6}}$

$\underline{\underline{[N_2H_4] = 0.11 \approx \text{Molarity Hydrazine Solution}}}$

14.40 $H_3PO_4 = H^+ + H_2PO_4^-$ $H_2PO_4^- = H^+ + HPO_4^{2-}$

$HPO_4^{2-} = H^+ + PO_4^{3-}$ $K_1 = \frac{[H^+][H_2PO_4^-]}{[H_3PO_4]}$ $K_2 = \frac{[H^+][HPO_4^{2-}]}{[H_2PO_4^-]}$

$K_3 = \frac{[H^+][PO_4^{3-}]}{[HPO_4^{2-}]}$ $\underline{K_1 = 7.5\times10^{-3}}$, $\underline{K_2 = 6.2\times10^{-8}}$, $\underline{K_3 = 2.2\times10^{-12}}$

CONC	H_2O	ADD	ADJ1	ADJ2	ADJ3	EQ.	Assume: $\therefore$		
H_3PO_4	—	1.0	$-X$	—	—	$1.0-X$	$X \ll 1.0$	1.0	$X = [H^+]$
H^+	10^{-7}	—	$+X$	$+Y$	$+Z$	$10^{-7}+X+Y+Z$	$X \gg 10^{-7}$	X	$X = [H_2PO_4^-]$
$H_2PO_4^-$	—	—	$+X$	$-Y$	—	$X-Y$		X	$Y = [HPO_4^{2-}]$
HPO_4^{2-}	—	—	—	$+Y$	$-Z$	$Y-Z$	$Y \ll X$	Y	
PO_4^{3-}	—	—	—	—	$+Z$	Z	$Z \ll Y$	Z	$Z = [PO_4^{3-}]$

$7.5\times10^{-3} = \frac{(X)(X)}{1.0}$, $X^2 = 7.5\times10^{-3}$ $\underline{\underline{X = 8.7\times10^{-2}}}$ (NOTE: X is < 1.0)

$6.2\times10^{-8} = \frac{(X)(Y)}{X}$ $\therefore$ $\underline{\underline{Y = 6.2\times10^{-8} = [HPO_4^{2-}]}}$ (NOTE: Y is $\ll$ X)

$2.2\times10^{-12} = \frac{XZ}{Y}$ $\therefore \underline{\underline{Z}} = 2.2\times10^{-12}\times6.2\times10^{-8}/8.7\times10^{-2} = \underline{\underline{1.6\times10^{-18}}}$ $[PO_4^{3-}]$

(NOTE: Z is $\ll$ Y)

14.42 (1) $H_2CO_3 = H^+ + HCO_3^-$ (2) $HCO_3^- = H^+ + CO_3^{2-}$

$$K_1 = \frac{[H^+][HCO_3^-]}{[H_2CO_3]} = 4.3 \times 10^{-7} \qquad K_2 = \frac{[H^+][CO_3^{2-}]}{[HCO_3^-]} = 5.6 \times 10^{-11}$$

pH=3.00 ∴ $[H^+] = 10^{-3}$ Assume: 0.1M → $[H_2CO_3] = 0.1$

$$4.3 \times 10^{-7} = \frac{(10^{-3})[HCO_3^-]}{0.1} \quad \therefore [HCO_3^-] = 4.3 \times 10^{-5}$$

$$5.6 \times 10^{-11} = \frac{(10^{-3})[CO_3^{2-}]}{4.3 \times 10^{-5}} \quad \therefore [CO_3^{2-}] = 2.4 \times 10^{-12}$$

(Sat. $H_2S = 10^{-1}$ M) pH=4.60

14.44 $H_2S = 2H^+ + S^{2-}$ $\quad K_1 \times K_2 = \frac{[H^+]^2[S^{2-}]}{[H_2S]} = 1.1 \times 10^{-21}$

$[H^+] = 2.5 \times 10^{-5}$ ∴ $[S^{2-}] = 1.1 \times 10^{-21}(10^{-1})/(2.5 \times 10^{-5})^2$ $\quad [S^{2-}] = 1.8 \times 10^{-13}$

14.46 (a) $1.8 \times 10^{-5} = \frac{[NH_4^+][OH^-]}{[NH_3]} = \frac{(0.10)[OH^-]}{(0.10)}$ ∴ $[OH^-] = 1.8 \times 10^{-5}$

pH = 9.26

(b) $1.8 \times 10^{-5} = \frac{[H^+][C_2H_3O_2^-]}{[HC_2H_3O_2]} = \frac{[H^+](0.40)}{(0.20)}$ ∴ $[H^+] = 9.0 \times 10^{-6}$

pH=5.05

(C) 8.29=pH (d) CAREFUL! HCl is a strong Acid! $[H^+] = 0.2$ pH= 0.7

14.48 $1.8 \times 10^{-5} = \frac{[OH^-][NH_4^+]}{[NH_3]}$ $\quad$ pH = 10.0 ∴ $[H^+] = 10^{-10}$, $[OH^-] = 10^{-4}$

$$\frac{(10^{-4})[NH_4^+]}{[NH_3]} = 1.8 \times 10^{-5} \quad \therefore \frac{[NH_4^+]}{[NH_3]} = 1.8 \times 10^{-1} \quad \frac{[NH_3]}{[NH_4^+]} = 5.6$$

14.50 (1) $HCHO_2 = H^+ + CHO_2^-$ $\quad K_a = \dfrac{[H^+][CHO_2^-]}{[HCHO_2]} = 1.8\times10^{-4}$

Conc.	H_2O	ADD(1)	ADD(2)	ADD(3)	ADJ.	EQ	∴	
$HCHO_2$	—	0.45	—	—	+X	.45+X	Assume	0.55
H^+	10^{-7}	—	—	0.10	−X	.10−X	$X\approx0.1$	$[H^+]$
CHO_2^-	—	—	0.55	—	−X	.55−X		0.45

$\dfrac{[H^+](0.45)}{(0.55)} = 1.8\times10^{-4}$ $\quad [H^+]=2.2\times10^{-4}$ $\quad pH=3.66$ $\quad$ (Before ADD(3) $[H^+]=1.47\times10^{-4}$ $pH=3.83$)

$\Delta pH = 0.17$

NOTE: $0.10-X=[H^+]$ $[H^+]=1.47\times10^{-4}$ ∴ $X\approx0.1$

14.54 (1) $A^- + H_2O = HA + OH^-$ $\quad K_{hy} = \dfrac{[HA][OH^-]}{[A^-]} = \dfrac{10^{-14}}{K_a}$

0.1 M NaA → $[A^-]\approx0.10$ $\quad pH=9.35$

∴ $[H^+]=4.47\times10^{-10}$ ∴ $[OH^-]=2.24\times10^{-5}$, $\quad \dfrac{K_a}{10^{-14}} = \dfrac{[A^-]}{[HA][OH^-]}$ $\quad$ ($[OH^-]\approx[HA]$)

$K_a = \dfrac{(10^{-14})(10^{-1})}{(2.24\times10^{-5})^2} = 2.0\times10^{-6}$

14.60 $Na^+B_3^- + H_2O = Na^+ + HB_3 + OH^-$ $\quad K_{hy} = \dfrac{[HB_3][OH^-]}{[B_3^-]} = \dfrac{10^{-14}}{6.6\times10^{-5}}$

CONC.	H_2O	ADD	ADJ.	EQ.	∴	
B_3^-	—	0.020	−X	0.020−X	Assume:	0.020
HB_3	—	—	+X	X	$X\ll0.02$	X
OH^-	10^{-7}	—	+X	$10^{-7}+X$	$X\gg10^{-7}$	X

$K_{hy}=1.5\times10^{-10}$

$1.5\times10^{-10} = \dfrac{(X)(X)}{0.020}$ $\quad X^2=3.0\times10^{-12}$ $\quad X=1.7\times10^{-6}$

NOTE: Assumptions valid

$[OH^-]=1.7\times10^{-6}$, $pOH=5.77$ ∴ $pH=8.23$

15.12 $PbCO_3 \rightleftharpoons Pb^{2+} + CO_3^{2-}$ $[Pb^{2+}]=[CO_3^{2-}]= 1.8\times10^{-7}$

$K_{sp}=[Pb^{2+}][CO_3^{2-}]=(1.8\times10^{-7})^2=\underline{\underline{3.2\times10^{-14}}}$

15.14 $CaCrO_4 = Ca^{2+} + CrO_4^{2-}$ $K_{sp}=[Ca^{2+}][CrO_4^{2-}]$

$[Ca^{2+}]=[CrO_4^{2-}]=\underline{1.0\times10^{-2}}$ $\therefore \underline{K_{sp}}=(1.0\times10^{-2})^2=\underline{\underline{1.0\times10^{-4}}}$

15.18 $Bi_2S_3 = 2Bi^{3+} + 3S^{2-}$ $K_{sp}=[Bi^{3+}]^2[S^{2-}]^3$

$$\frac{?\,\text{mol } Bi_2S_3}{1\,\ell\text{ sol}}=\frac{2.5\times10^{-12}\text{ g}}{1\,\ell\text{ sol}}\times\frac{1\text{ mol } Bi_2S_3}{514.18\text{ g } Bi_2S_3}=\underline{4.86\times10^{-15}\,M}$$

$\underline{[Bi^{3+}]=2\times(M)}$, $\underline{[S^{2-}]=3\times(M)}$ $\therefore K_{sp}=(9.72\times10^{-15})^2(1.46\times10^{-14})^3$

$\underline{\underline{K_{sp}=2.9\times10^{-70}}}$

15.20 (a) $PbS = Pb^{2+} + S^{2-}$ $K_{sp}=[Pb^{2+}][S^{2-}]=7\times10^{-27}$

$\underline{(M)}=[Pb^{2+}]=[S^{2-}]=\sqrt{7\times10^{-27}}=\underline{\underline{8\times10^{-14}}}$

(b) $\underline{\underline{7.9\times10^{-6}}}$ (c) $\underline{\underline{3.9\times10^{-5}}}$ (d) $\underline{Hg_2Cl_2} = Hg_2^{2+} + 2Cl^-$ $K_{sp}=[Hg_2^{2+}][Cl^-]^2$

$K_{sp}=2\times10^{-18}$

$[Cl^-]=2[Hg_2^{2+}]=2(M)$, $2\times10^{-18}=(M)(2M)^2=4(M)^3$

$\underline{\underline{(M)=8\times10^{-7}}}$

(e) $\underline{Al(OH)_3}=Al^{3+}+3OH^-$, $K_{sp}=[Al^{3+}][OH^-]^3$

$[OH^-]=3[Al^{3+}]=3(M)$, $K_{sp}=2\times10^{-33}=(M)(3M)^3=27(M)^4$

$(M)=3\times10^{-9}$ (f) 9×10^{-3}

15.22 $CaSO_4 \rightleftharpoons Ca^{2+} + SO_4^{2-}$ $K_{sp}=2\times10^{-4}=[Ca^{2+}][SO_4^{2-}]$

$[Ca^{2+}]=[SO_4^{2-}]=\text{Formality}=\sqrt{2\times10^{-4}}=\underline{1.4\times10^{-2}\text{ mol }\ell^{-1}}$

$$?\text{ g } CaSO_4 = 600\text{ ml}\times\frac{1.4\times10^{-2}\text{ mol}}{1000\text{ ml}}\times\frac{136\text{ g } CaSO_4}{1\text{ mol } CaSO_4}=\underline{\underline{1\text{ g } CaSO_4}}$$

(assume vol. of water ≈ vol. of sol.)

NOTE: only one sig. fig. — SEE K_{sp} value

15.24 $CaCO_3 = Ca^{2+} + CO_3^{2-}$ $K_{SP} = [Ca^{2+}][CO_3^{2-}] = 9 \times 10^{-9}$
Let X = molar solubility of $CaCO_3$ $\therefore [Ca^{2+}] = X$, $[CO_3^{2-}] = 0.50 + X$
Assume: $X \ll 0.50$ $\therefore [CO_3^{2-}] \approx 0.50$ $0.50X = 9 \times 10^{-9}$, $X = 1.8 \times 10^{-8}$

15.28 $CaF_2 = Ca^{2+} + 2F^-$ $K_{SP} = [Ca^{2+}][F^-]^2 = 1.7 \times 10^{-10}$
$[Ca^{2+}] = (M)$ $\therefore [F^-] = 2(M) + 0.010$ Assume: $2(M) \ll 0.010$
$\therefore [F^-] \approx 0.010$ $\therefore (M)(0.010)^2 = 1.7 \times 10^{-10}$ $(M) = 1.7 \times 10^{-6}$

15.30 $AgC_2H_3O_2 = Ag^+ + C_2H_3O_2^-$ $K_{SP} = [Ag^+][C_2H_3O_2^-] = 2.3 \times 10^{-3}$
$[Ag^+] = 0.05$, $[C_2H_3O_2^-] = 0.001$ ion product $= 5 \times 10^{-5} < K_{SP}$
$\therefore$ No ppt. (b) $K_{SP} = [Ba^{2+}][F^-]^2 = 1.7 \times 10^{-6}$ $[Ba^{2+}] = 1.0 \times 10^{-2}$
$[F^-] = 2.0 \times 10^{-2}$ ion product $= (2.0 \times 10^{-2})^2 (1.0 \times 10^{-2}) = 4.0 \times 10^{-6}$
$> K_{SP}$ $\therefore$ ppt. forms (C) $K_{SP} = 2 \times 10^{-4} = [Ca^{2+}][SO_4^{2-}]$
$[Ca^{2+}] = 500/750 \times 1.4 \times 10^{-2}$, $[SO_4^{2-}] = 250/750 \times 0.25$
ion prod. $= 7.8 \times 10^{-4} > 2 \times 10^{-4}$ $\therefore$ ppt. forms

15.32 $AgCl = Ag^+ + Cl^-$ $K_{SP} = [Ag^+][Cl^-] = 1.7 \times 10^{-10}$

CONC.	initial	after mix	ADJ.	EQ	Assume	$\therefore$
Ag^+	0.20	0.10	$-X$	$0.10-X$	$X \approx 0.050$	0.050
Cl^-	0.10	0.050	$-X$	$0.050-X$		very small

$0.050[Cl^-] = 1.7 \times 10^{-10}$ $\therefore [Cl^-] = 3.4 \times 10^{-9}$ (assump. valid)
$[Ag^+] = 0.050$, $[NO_3^-] = 0.10$, $[H^+] = 0.050$

15.34 $K_{SP}(PbCrO_4) = 1.8 \times 10^{-14}$ $K_{SP}(BaCrO_4) = 2.4 \times 10^{-10}$
to ppt $PbCrO_4$: $[CrO_4^{2-}] = 1.8 \times 10^{-12}$ to ppt $BaCrO_4$: $[CrO_4^{2-}] = 2.4 \times 10^{-8}$
$\therefore PbCrO_4$ first $[Pb^{2+}] = 7.5 \times 10^{-7}$ when ppt. $BaCrO_4$ starts

15.36 $K_{sp} = [Fe^{2+}][S^{2-}] = 3.7 \times 10^{-19}$, $K_{sp} = [Zn^{2+}][S^{2+}] = 1.2 \times 10^{-23}$

$0.10\,[S^{2-}] = 3.7 \times 10^{-19}$, $[S^{2-}] = 3.7 \times 10^{-18}$ ← Largest $[S^{2-}]$ which will leave all (0.10) Fe^{2+} in solution

$$K_1 \times K_2 = \frac{[H^+]^2[S^{2-}]}{[H_2S]} = 1.1 \times 10^{-21}$$

Saturated $H_2S \rightarrow [H_2S] = 0.10$

$$\therefore \frac{[H^+]^2(3.7 \times 10^{-18})}{(0.10)} = 1.1 \times 10^{-21} \quad \therefore [H^+] \geq 5.5 \times 10^{-3}\,M \;(< 0.96\,M)$$

$[Zn^{2+}](3.7 \times 10^{-18}) = 1.2 \times 10^{-23}$ $\therefore [Zn^{2+}] = 3.2 \times 10^{-6}\,M$

15.38 $K_{sp}(ZnS) = 1.2 \times 10^{-23} = [Zn^{2+}][S^{2-}]$ $[H^+] = 12$

$$K_1 \times K_2 = \frac{[H^+]^2[S^{2-}]}{[H_2S]} = 1.1 \times 10^{-7} \times 1.0 \times 10^{-14}$$

$[H_2S]$ (limit) = 10^{-1} (solubility)

$\therefore [S^{2-}]$ (limit) $= 10^{-1} \times 1.1 \times 10^{-21}/(12)^2 = 7.6 \times 10^{-25}$

$[Zn^{2+}]$(limit) $= 1.2 \times 10^{-23}/[S^{2-}]$(limit) $= 16$ ← This value is larger than necessary to permit a ZnS ppt. to dissolve.

15.40 $Zn(OH)_2 = Zn^{2+} + 2OH^-$ $K_{sp} = [Zn^{2+}][OH^-]^2 = 4.5 \times 10^{-17}$

$[OH^-] = 2 \times (M) = 2 \times 5.7 \times 10^{-3} = 1.14 \times 10^{-2}$ $\therefore [Zn^{2+}](1.14 \times 10^{-2})^2 = 4.5 \times 10^{-17}$

$[Zn^{2+}] = 3.5 \times 10^{-13}$ $Zn(NH_3)_4^{2+} = Zn^{2+} + 4\,NH_3$ (most of Zn^{2+} is $Zn(NH_3)_4^{2+}$)

$$K_{inst.} = \frac{[Zn^{2+}][NH_3]^4}{[Zn(NH_3)_4^{2+}]} = \frac{(3.5 \times 10^{-13})(1.0)^4}{(5.7 \times 10^{-3})} = 6.1 \times 10^{-11}$$

CHAPTER 16

16.32 $$?e = 1\ \text{coul} \times \frac{1\ F}{96{,}500\ \text{coul}} \times \frac{6.022 \times 10^{23}\ e}{1\ F} = 6.24 \times 10^{18}\ e$$

16.34 (a) $$?F = 8950\ \text{coul} \times \frac{1\ F}{96{,}500\ \text{coul}} = 0.0927\ F$$

(b) $$?F = 1.5\ \text{amp} \times 30\ \text{sec} \times \frac{1\ \text{coul sec}^{-1}}{1\ \text{amp}} \times \frac{1\ F}{96{,}500\ \text{coul}} = 4.66 \times 10^{-4}\ F$$

(c) $9.14 \times 10^{-2}\ F$

16.36 (a) $$?\text{min} = \frac{84{,}200\ \text{coul}}{6.30\ \text{amp}} \times \frac{1\ \text{amp}}{1\ \text{coul sec}^{-1}} \times \frac{1\ \text{min}}{60\ \text{sec}} = 223\ \text{min}$$

(b) 239 min

(c) $$?\text{min} = \frac{0.50\ \text{mol Al}}{18.3\ \text{amp}} \times \frac{3 \times 96{,}500\ \text{coul}}{1\ \text{mol}\ Al^{3+}} \times \frac{1\ \text{amp}}{1\ \text{coul sec}^{-1}} \times \frac{1\ \text{min}}{60\ \text{sec}}$$
$= 132\ \text{min}$

16.38 $$?\text{g Na} = 25\ \text{amp} \times 8\ \text{hr} \times \frac{3600\ \text{sec}}{1\ \text{hr}} \times \frac{1\ \text{coul sec}^{-1}}{1\ \text{amp}} \times \frac{23.0\ \text{g Na}}{96{,}500\ \text{coul}}$$
$= 172\ \text{g Na}$ ($265\ \text{g}\ Cl_2$)

16.40 $$?\text{g Cu} = 100\ \text{amp} \times 8\ \text{hr} \times \frac{1\ \text{coul sec}^{-1}}{1\ \text{amp}} \times \frac{3600\ \text{sec}}{1\ \text{hr}} \times \frac{63.54\ \text{g Cu}}{2 \times 96{,}500\ \text{coul}} =$$
948 g Cu (900 to one Sig. Fig.)

16.42 $$?\text{sec} = \frac{21.4\ \text{g Ag}}{10.0\ \text{amp}} \times \frac{1\ \text{amp}}{1\ \text{coul sec}^{-1}} \times \frac{96{,}500\ \text{coul}}{107.9\ \text{g Ag}} = 1910\ \text{sec}$$

16.44 $? \text{min} = \dfrac{5 \text{ g Cu}}{5 \text{ amp}} \times \dfrac{1 \text{ amp}}{1 \text{ coul sec}^{-1}} \times \dfrac{2 \times 96{,}500 \text{ coul}}{63.5 \text{ g Cu}} \times \dfrac{1 \text{ min}}{60 \text{ sec}} = \underline{\underline{50.66 \text{ min}}}$

(50 min to 1 S.F.)

16.46 $? \text{amp} = \dfrac{1.33 \text{ g } Cl_2}{45 \text{ min}} \times \dfrac{96{,}500 \text{ coul}}{35.45 \text{ g } Cl_2} \times \dfrac{1 \text{ min}}{60 \text{ sec}} \times \dfrac{1 \text{ amp}}{1 \text{ coul sec}^{-1}}$

$= \underline{\underline{1.34 \text{ amp}}}$

16.48 $? \text{mol Cr} = 0.125 \text{ mol Cu} \times \dfrac{2 \mathcal{F}}{1 \text{ mol Cu}} \times \dfrac{1 \text{ mol Cr}}{3 \mathcal{F}} = \underline{\underline{0.0833 \text{ mol Cr}}}$

16.52 (a) $Al \rightarrow Al^{3+} + 3e$ (1.71) $Ni^{2+} \rightarrow Ni - 2e$ (−0.25) $\underline{\underline{\mathcal{E}^\circ = 1.46}}$

(b) $PbO_2 + SO_4^{2-} + 4H^+ + 2e \rightarrow PbSO_4 + 2H_2O$ (1.69)

$2Cr^{3+} + 7H_2O \rightarrow Cr_2O_7^{2-} + 14H^+ + 6e$ (−1.33) $\underline{\underline{\mathcal{E}^\circ = 0.36}}$

(c) $Ag^+ \rightarrow Ag - 1e$ (0.80) $Pb \rightarrow Pb^{2+} + 2e$ (0.13) $\underline{\underline{\mathcal{E}^\circ = 0.93}}$

(d) $Cl_2 \rightarrow 2Cl^- - 2e$ (1.36), $Mn^{2+} + 2H_2O \rightarrow MnO_2 + 4H^+ + 2e$ (−1.28)

$\underline{\underline{\mathcal{E}^\circ = 0.08}}$ (e) $Mn \rightarrow Mn^{2+} + 2e$ (1.03), $2H^+ \rightarrow H_2 - 2e$ (0.00) $\underline{\underline{\mathcal{E}^\circ = 1.03}}$

16.54 (a) $K_{EQ} = [Ni^{2+}]/[Sn^{2+}]$ $\Delta G = \Delta G^\circ + RT \ln (\text{M.A.})$

at EQ. $\Delta G = 0$, $(\text{M.A.}) = K_{EQ}$ $\therefore \Delta G^\circ = -RT \ln K_{EQ}$ and $\Delta G^\circ = -n\mathcal{F}\mathcal{E}^\circ$

$\therefore \ln K_{EQ} = n\mathcal{F}/RT \; \mathcal{E}^\circ$ $K_{EQ} = e^{(n\mathcal{F}/RT)\mathcal{E}^\circ}$

$\mathcal{E}^\circ = 0.25 - 0.14 = \underline{0.11}$, $\underline{n = 2}$, $\underline{RT/\mathcal{F} = 0.0257 \text{ J coul}^{-1}}$ $\underline{\underline{K_{EQ} = 5.2 \times 10^3}}$

(b) $\underline{K_{EQ} = 10^{(2/0.0592)\mathcal{E}^\circ}}$ $\underline{\mathcal{E}^\circ = 1.36 - 1.09 = 0.27}$ $\underline{\underline{K_{EQ} = 1.3 \times 10^9}}$

(c) $\underline{\mathcal{E}^\circ = 0.80 - 0.77} = \underline{0.03}$ $\underline{\underline{K_{EQ} = 3}}$

16.56 $\Delta G° = -n\mathcal{F}\mathcal{E}° = -RT \ln K_{EQ}$ $\therefore K_{EQ} = e^{n\mathcal{F}\mathcal{E}°/RT}$ (assume 25°C)

$\therefore \mathcal{F}/RT = 96{,}500\ \text{coul}/8.314\ \text{J}\ °\text{K}^{-1}\text{mol}^{-1} \times 298°\text{K} = \underline{38.95\ \text{coul J}^{-1}\text{mol}^{-1}}$

since $1\text{J} = 1\text{volt} \times 1\text{coul}$ $\mathcal{F}/RT = \underline{38.95\ \text{volt}^{-1}\text{mol}^{-1}}$

$K_{EQ} = e^{n\mathcal{E}° \times 38.95\ \text{volt}^{-1}\text{mol}^{-1}}$ $\mathcal{E}°_{(a)} = -2.76\ \text{volt} + 2.38\ \text{volt} = \underline{-0.38\ \text{volt}}$

$n_{(a)} = \underline{2}$ $\therefore K_{EQ} = e^{-29.60} = \underline{\underline{1.39 \times 10^{-13}}}$

(b) $\mathcal{E}° = \underline{-1.49\ \text{volt}}$, $K_{EQ} = \underline{\underline{3.90 \times 10^{-51}}}$ (c) $\mathcal{E}° = \underline{0.64\ \text{volt}}$, $K_{EQ} = \underline{\underline{4.49 \times 10^{21}}}$

(d) $\mathcal{E}° = \underline{-0.16\ \text{volt}}$, $K_{EQ} = \underline{\underline{6.37 \times 10^{-82}\ (n=30)}}$ (e) $\mathcal{E}° = \underline{-0.13}$, $K_{EQ} = \underline{\underline{1.60 \times 10^{-9}}}$

16.58 (a) $\Delta G°_{298} = -n\mathcal{F}\mathcal{E}° = -2 \times 96{,}500\ \mathcal{E}°$ $\underline{\mathcal{E}° = 2.38 - 2.76 = -0.38}$

$\underline{\underline{\Delta G° = 7.33 \times 10^4\ \text{J}}}$ (b) $\underline{\mathcal{E}° = -0.13 - 1.36 = -1.49}$ $\therefore \underline{\underline{\Delta G° = 2.88 \times 10^5\ \text{J}}}$

(c) $\underline{\mathcal{E}° = 2.00 - 1.36 = 0.64}$ $\therefore \underline{\underline{\Delta G° = -1.24 \times 10^5\ \text{J}}}$

(d) $\underline{\mathcal{E}° = 1.33 - 1.49 = -0.16}$ $\therefore \underline{\underline{\Delta G° = 4.64 \times 10^5}}$

(e) $\underline{\mathcal{E}° = 1.23 - 1.36 = -0.13}$ $\therefore \underline{\underline{\Delta G° = 5.02 \times 10^4\ \text{J}}}$

16.60 (a) $\mathcal{E}° = 1.71 - 0.25 = \underline{\underline{1.46\ \text{volts}}}$ $\mathcal{E} = \mathcal{E}° - RT/n\mathcal{F}\ \ln [Al^{3+}]^2/[Ni^{2+}]^3$

$\mathcal{E} = 1.46\ \text{volt} - 0.0592/6\ \text{volt} \times \log (0.02)^2/(0.80)^2 = \underline{\underline{1.49\ \text{volts}}}$

$\Delta G = -n\mathcal{F}\mathcal{E} = -(6)(96{,}500\ \text{coul})(1.49\ \text{volt}) \times 1\text{J}/1\text{coul} \times 1\text{volt} = \underline{\underline{-8.63 \times 10^5\ \text{J}}}$

(b) $\mathcal{E}° = \underline{\underline{0.11\ \text{volt}}}$, $\mathcal{E} = \underline{\underline{0.17\ \text{volt}}}$, $\Delta G = \underline{\underline{-3.29 \times 10^4\ \text{J}}}$

(c) $\mathcal{E}° = \underline{\underline{1.28\ \text{volt}}}$, $\mathcal{E} = \underline{\underline{1.26\ \text{volt}}}$, $\Delta G = \underline{\underline{-2.44 \times 10^5\ \text{J}}}$ (actually Cu^+ disprop. in aq. solution)

16.62 $\mathcal{E} = \mathcal{E}° - \frac{RT}{n\mathcal{F}} \ln(\text{M.A.})$ (for a conc. cell $\mathcal{E}° = 0$) $\therefore \mathcal{E} = -\frac{RT}{n\mathcal{F}} \ln(\text{M.A.})$

$\mathcal{E} = -\frac{0.0592}{n} \log(\text{M.A.})$ $\underline{(\text{M.A.}) = [Fe^{2+}]_F/[Fe^{2+}]_I = 10^{-3}/10^{-1}}$ $\underline{(\text{M.A.}) = 10^{-2}}$

$\underline{\underline{\mathcal{E}}} = -0.030 \log 10^{-2} = \underline{\underline{0.06\ \text{Volt}}}$

16.64 $\mathcal{E} = \mathcal{E}^\circ - RT/n\mathcal{F} \ln(M.A.)$ $Ag^+(Br^-) + \frac{1}{2}H_2 \rightarrow Ag + (Br^-) + H^+$

$\mathcal{E} = 0.80 - 0.0592/1 \times \log [H^+]/[Ag^+]$ since $K_{SP}(AgBr) = \underline{5 \times 10^{-13}}$ and $[Br^-] = \underline{10^{-2}}$ $[Ag^+] = \underline{5 \times 10^{-11}}$, $[H^+] = \underline{1.0}$ $\therefore \log 1/5 \times 10^{-11} = \underline{10.3}$ $\mathcal{E} = \underline{\underline{0.19}}$

16.66 $\mathcal{E} = \mathcal{E}^\circ - (RT/n\mathcal{F}) \ln(M.A.)$ $Pb \rightarrow Pb^{2+} + 2e$ $(M.A.) = [Pb^{2+}]$

$\mathcal{E}^\circ - \mathcal{E} = 0.030 \log[Pb^{2+}]$ $0.13 - 0.51 = 0.030 \log[Pb^{2+}] = -0.38$

$\log[Pb^{2+}] = -0.38/0.03$ $\underline{[Pb^{2+}] = 10^{-12.7}} = \underline{2.15 \times 10^{-13}}$, $\underline{[CrO_4^{2-}] = 0.10}$

$\underline{\underline{K_{SP}(PbCrO_4)}} = [Pb^{2+}][CrO_4^{2-}] = \underline{\underline{2.15 \times 10^{-14}}}$

18.44 $Na_{(s)} \rightarrow Na^{2+}_{(aq)} + 2e$ (overall process) ΔH°_{OA} (to be negative)

$Na_{(s)} \rightarrow Na_{(g)}$ $\Delta H^\circ_{atom} = 25.98\ kcal\ mol^{-1}$

$Na_{(g)} \rightarrow Na^{+}_{(g)} + 1e$ $I.E._{(1)} = 118.0\ kcal\ mol^{-1}$ } $1228\ kcal\ mol^{-1}$

$Na^{+}_{(g)} \rightarrow Na^{2+}_{(g)} + 1e$ $I.E._{(2)} = 1084\ kcal\ mol^{-1}$

$Na^{2+}_{(g)} \rightarrow Na^{2+}_{(aq)}$ $\Delta H^\circ_{hyd} = ?$ Would be larger than $-1228 \frac{kcal}{mol}$

18.46 $\Delta G^\circ = -RT \ln K_p$ $\Delta G^\circ = \Delta H^\circ - T\Delta S^\circ$ If $K_p = 1$ $\therefore \Delta G^\circ = 0$

$\therefore \Delta H^\circ = T\Delta S^\circ$ $\Delta H^\circ = 155\ kJ\ mol^{-1}$ $\Delta S^\circ = S^\circ_{Cu(s)} + \frac{1}{2}S^\circ_{O_2(g)} - S^\circ_{CuO(s)}$

$T = \Delta H^\circ / \Delta S^\circ$ $\Delta S^\circ = 33.3\ J(mol{\cdot}K)^{-1} + \frac{1}{2} \times 205.0\ J(mol{\cdot}K)^{-1} - 43.5\ J(mol{\cdot}K)^{-1}$

$\Delta S^\circ = 92.3\ J\ mol^{-1}K^{-1}$ $\therefore$ $T = 1.55 \times 10^5\ J\ mol^{-1} / (92.3\ J\ mol^{-1}K^{-1})^{-1} = 1679\ K$

(1680 K to 3 sig. fig.)

CHAPTER 20

20.54 Zr = 1.60 Å Hf = 1.58 Å (hcp = same density as fcc)

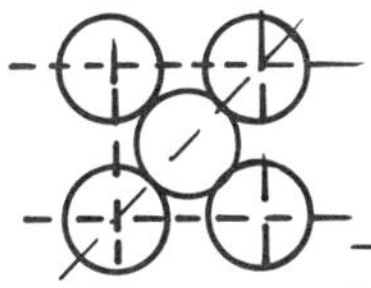

Note Unit Cell face diagonal = 4 radii (4 atoms/u.c.)

Edge = face diag. ÷ $\sqrt{2}$ Volume = $(\text{Edge})^3$

Zr = $(4 \times 1.60\,\text{Å} \times 2^{-\frac{1}{2}})^3$ = 92.68 Å³ Hf = 89.25 Å³

$$\frac{?\,\text{g Hf}}{1\,\text{cm}^3\,\text{Hf}} = \frac{6.49\,\text{g Zr}}{1\,\text{cm}^3\,\text{Zr}} \times \frac{92.68\,\text{Å}^3\,\text{Zr}}{1\,\text{unit cell}} \times \frac{6.022 \times 10^{23}\,\text{at}}{91.22\,\text{g Zr}} \times \frac{1\,\text{unit cell}}{89.25\,\text{Å}^3\,\text{Hf}} \times \frac{178.5\,\text{g Hf}}{N\,\text{at}}$$

= 13.2 g Hf/1 cm³ Hf

23.26 fraction remaining $= 1/2^n$ (n = no. of half-life periods)

(a) $1g \times 1/2^1 = \underline{\underline{0.500 g}}$ (b) $1g \times 1/2^3 = \underline{\underline{0.125 g}}$ (c) $\underline{\underline{0.0313 g}}$

23.28

$$t_{\frac{1}{2}} = \frac{0.693}{k} = \frac{0.693}{4.23 \times 10^{-3}\ \text{days}^{-1}} = \underline{\underline{164\ \text{days}}}$$

23.30

$$\log\frac{[^{51}Cr]_o}{[^{51}Cr]_t} = \frac{kt}{2.303} = \log 2 \ (\text{when } t = 27.72 \text{ days})$$

$$\therefore k = \frac{2.303 \times \log 2}{27.72\ \text{days}} = \underline{\underline{2.52 \times 10^{-2}\ \text{days}^{-1}}}$$

23.32 mol ^{40}Ar (formed) = mol ^{40}K (decayed) $= 1.15 \times 10^{-5}$ mol

$$k = \frac{0.693}{t_{\frac{1}{2}}} = \frac{0.693}{1.3 \times 10^9\ \text{yr}} = \underline{5.3 \times 10^{-10}\ \text{yr}^{-1}} \qquad t = \frac{2.303}{k} \times \log\frac{[A]_o}{[A]_t}$$

$[A]_o = \underline{(2.07 \times 10^{-5} + 1.15 \times 10^{-5})\ \text{mol}\ ^{40}K}$ $\quad [A]_t = \underline{2.07 \times 10^{-5}\ \text{mol}\ ^{40}K}$

$$\therefore t = \frac{2.303}{5.3 \times 10^{-10}\ \text{yr}^{-1}} \times \log\frac{3.22 \times 10^{-5}}{2.07 \times 10^{-5}} = \underline{\underline{8.3 \times 10^8\ \text{yr}}}$$

23.34 $E = mc^2$ $\quad (J = kg \times (m\ sec^{-1})^2 \quad \therefore E = 1.82 \times 10^{-30} kg (2.998 \times 10^8 m)^2$

$\underline{\underline{E = 1.64 \times 10^{-13} J}}$